AVIATRIX

by Elinor Smith

Thorndike Press • Thorndike, Maine

Library of Congress Cataloging in Publication Data:

Smith, Elinor.
  Aviatrix.

  Originally published: New York : Harcourt Brace
Jovanovich, c1981.
  1. Smith, Elinor. 2. Air pilots – United States –
Biography. I. Title.
[TL540.S64A3 1982]   629.13'092'4 [B]   82-5849
ISBN 0-89621-368-4   AACR2

Large Print edition available through arrangement with
Harcourt Brace Jovanovich, Inc.

Cover design by Andy Winther.

To Harriet — who never doubted

Preface

I had been brought up to think that anyone could do anything he or she put his or her mind to, so I was shocked to learn that the world had stereotypes it didn't want tampered with. In an age when girls were supposed to be seen and not heard, look beautiful, and occasionally faint, I didn't seem to fit in anywhere.

Luckily I was used to being seen and not heard. The better part of my childhood in the 1920s was spent around the early planes and pioneer fliers at Roosevelt Field on Long Island, where nobody paid much attention to me anyhow. But being beautiful was something else again. My usual costume consisted of an old pair of my brother's knickers, a sport shirt, argyle socks, and a beat-up leather jacket. A cotton helmet, goggles, and sneakers were the finishing touches to my daily attire, and I defy anyone to be alluring in that getup, even without my rich crop of orangy freckles. As for the genteel fainting spells, they didn't jibe at all with my consuming ambition to fly airplanes for hire.

Unfortunately my male colleagues summed up my prospects for a career as a professional pilot in one word: bleak. They felt that if they — with far more flying hours to their credit than I had — were having such a difficult time making a living, what chance did a teen-age girl stand? However, I had a distinct advantage over them. I was the only teenage girl in the country possessing a pilot's license. I argued that the publicity resulting from this accomplishment made my prospects for success much brighter.

Besides, my father's optimism about flying was contagious, and I had long since succumbed to his noting that the airplane would easily outstrip the discovery of fire and the invention of the wheel in its value to humanity. If the rumors I heard were correct, the Guggenheims were providing a fund of several million dollars to be devoted to a study of flight under adverse weather conditions. This meant that the development of aircraft instrumentation would now be more than a distant dream, particularly because the pilot of this ambitious project was to be none other than my own personal hero, Lieutenant James H. Doolittle. Clearly this was the time to hang in there and await developments.

A great deal of the excitement of the transatlantic flight craze in 1927 — when "Lucky Lindy" flew to Paris and was quickly followed by Clarence Chamberlin and Commander Byrd — had rubbed off on me, and I wanted to become a part of it. After all, I reasoned, I could easily outfly Ruth Elder, the first female to brave the ocean in 1927, and, as things turned out, the second one, too. For Amelia Earhart openly admitted that she had only been a passenger aboard the *Friendship* when it made its crossing in 1928.

In that same year, 1928, I first attracted world-wide attention by flying a landplane under New York's four East River bridges, something I don't believe has been done since. It was a very difficult feat at the time, and this book tells the true story behind that unprecedented flight and many other trials and adventures my colleagues and I had in the early days of aviation.

I give many talks before college groups these

days and have found that students are always surprised to learn of the many female pilots in the twenties and thirties and their outstanding accomplishments. Their lack of information seems to be largely due to the fact that only Amelia Earhart's name has been widely publicized since that era. Yet there were many years immediately following her 1928 flight when she was far from being the "Lady Lindy" she was publicized as. This account will do much to clarify the Earhart "mystique" so carefully contrived by George Palmer Putnam, her manager and later her husband. Ironically enough, Putnam's machinations on Amelia's behalf led to my being named "Best Woman Pilot in the United States" in 1930 at the tender age of nineteen, for I was forced to set many different kinds of aviation records in order to circumvent his spheres of influence.

As for Amelia herself, I believe I knew her as well as, or better than, most. She was a brave lady, and what I relate here in no way diminishes her accomplishments. But I would also like to record and give credit to my other gallant friends whose names are largely unknown to the public today: Louise Thaden, Lady Mary Heath, Phoebe Omlie, Jessie "Chubby" Miller, "Pancho" Barnes, Laura Ingalls, Mae Haizlip, Gladys O'Donnell, and many others. Since the airlines and the military are finally letting down the bars to admit qualified young women, this is a good time to recall the difficulties that most women fliers encountered during our early struggles for recognition and employment.

Why did we persist in a business that offered so few financial rewards and took lives at such a cruel rate? It's a question that had as many answers as there were pilots. In my case it was the daily challenge and the sheer beauty of flight that drew me back again and again.

I still won't book a commercial flight unless I'm assured of a window seat, and as for in-flight movies, who needs Dustin Hoffman with all that glory and excitement swirling around outside the porthole?

Introduction

It was not until August 2, 1911, eight years after the Wright Brothers' historic flight at Kitty Hawk, that America licensed its first woman pilot – Harriet Quimby. By 1914, there were no more than a dozen women pilots, and it would take until the end of 1928 before that number rose to thirty-four. One of that number was Elinor Smith, who at sixteen became the youngest pilot to receive an FAI license; it was signed by Orville Wright.

Men and women pilots alike were considered a curious, impractical, daredevil breed until Charles Lindbergh's electrifying flight to France in May 1927. It seems very fitting to me that in that same month and year and from the same field a slim fifteen-year-old girl flew her first solo flight. And that girl was Elinor Patricia Smith.

Elinor Patricia Smith has always been known as Elinor — never as Patricia. Never "Pat". That tells a great deal about an active lady. She was sturdy and strong. She had to develop that strength, despite her diminutive size, because of the heavy demand on the muscles that flying in those days often required. Elinor wore grease-smeared overalls. Her blonde hair was often tucked, totally, beneath the leather headgear of a pilot. Her eyes could be merry, interested, pleasant, intent, farseeing, and often stern. But they were — and are — the color of the space she loves — bluest blue.

Elinor Smith had many avenues of life open to her. The daughter of a vaudeville headliner, the unforgettable Tom Smith, she could readily have found the Yellow Brick Road open, and with few obstacles. It's obvious she never even considered it. She only wanted to fly, and fly she did. She still does, for fun and for profit.

I have known Elinor for many years, from the days when she and "Slim" Lindbergh stirred up the same air over Long Island's Curtiss Field, later Roosevelt Field, now a land in chains to commerce — a shopping center. One of my earliest and most lasting impressions was when she touched down after that first women's altitude record. She was enraptured by her own accomplishment. She was near speechlessness – but not quite. She told us all about it, but there was never a trace of bragging. As always, Elinor was completely honest in her description and appraisal, and she has been that way ever since.

Elinor worked for what she accomplished, worked very hard indeed. Her mind was attuned to receive and develop certain crafts, but hard work was required to educate that mind, to whet the latent skills to the point of brilliant accomplishment. Elinor had the will power.

I have talked with her many times. Only now, after reading this volume in typescript, did I realize the many good stories she has carved out of that thin air at thousands of feet. She describes magnificent feats of flying, and her readers can feel the beginning of that pain in the back of the neck that comes from watching the breath-taking sight of a soaring, twisting gymnastic plane. Yet Elinor Smith could duplicate every one of those maneuvers easily – and in all probability, as well as or even more smoothly than the pilot she writes about.

She was born with another sense – the ability immediately to spot a phony, whether in flight goggles or a tailored suit. You'll see how she manages them in these pages.

She has rounded another pylon with this book. Now you can feel that you were there, as I was, some fifty years ago.

JOHN FROGGE

Acknowledgments

I am deeply indebted to the following friends for their unwavering encouragement and support:

Gene and Sue Bartczak, Eugene Husting, Lyn Doblemaier, Alvin Bahnsen, Andrew Yelaney, Shannon, Jake, and Mathew Seigle, Frances Streit, Thomas Bass, Diane Corday, Lois McFall and Alice Jackson.

To my children, Patrick, Patricia, Kathleen, and Pamela, my brother Joe, and their spouses Harriet, Marvin, and Dora, for their forbearance during what turned into a trying time.

To Charlotte Cassidy for her patience, tact, and competence as she typed up a pristine manuscript from a jumble of rewritten pages.

To my editor, Rhoda Schlamm, whose understanding and guidance far transcended the boundaries of official literary duties.

And to all of those good friends who aided me in placing times and dates correctly, thereby preventing me from leaning too hard on personal memory, a fragile reed at best: Slim Hennicke and Herb Ottewill of the Long Early Fliers; William Kaiser, Curator of the Cradle of Aviation Museum; Dick Winsche of the Museum in the Park, Long Island; Frank Strnad; John Underwood of Heritage Press; Claudia Oakes and Kathleen Brooks-Pazmany of the Smithsonian Institution; and Kenneth Foreman of the American Institute of Aeronautics and Astronautics; and the staff of the Library of the University of California at Santa Cruz, California.

John Frogge has been a top-flight newsman for more than half a century. In the mid-twenties, he was the *New York Times* reporter stationed on Roosevelt Field. He worked directly under their Aviation Managing Editor, Lauren D. (Deke) Lyman, in covering the 1927 Lindbergh flight. Subsequently, he moved over to the *New York Herald Tribune* and became their aviation reporter on Roosevelt Field. He watched the greats come and go as he plied his typewriter, remaining at the field almost to the day it became a shopping center in the fifties. His literary talents were highly respected by his colleagues, and his reporting integrity endeared him to all of the important members of the aviation community.

AVIATRIX

1

The View from the Bridges

I kicked the rudder hard right and laid the stick over to bring the wings level. The Waco was having a hard time riding out the turbulence created by brisk October breezes gusting in off the Atlantic and colliding abruptly with the heat rising from Brooklyn's closely packed buildings below.

From 800 feet, the view of New York's Lower Bay was spectacular. Sunshine glittered on the water, highlighting the white sails and dark hulls below, while casting bluish shadows against the columns of concrete and glass that formed New York's brilliant skyline. Over to my right the four bridges connecting Manhattan to Long Island were etched against the blue of the

East River like exquisitely fashioned Tinker-toys. The Statue of Liberty held her torch high, serenely contemplating the panorama below her like a goddess surveying her domain. I yearned for a small measure of her dignity and calm. Knowing that I was about to reduce that marine placidity to a chaotic shambles within the next ten minutes made my palms sweat in way I had never before experienced.

In 1928 flying under all the East River bridges in a landplane was no light-hearted pleasure hop. It was a slick piece of flying, requiring expert piloting and technical knowledge, certainly no job for a novice, let alone a seventeen-year-old girl. But I already had ten years of flying experience behind me, plus almost two years of solo time, much of it close to the water, so why was I getting the shakes now?

Well, for one thing, I'd painted myself into a corner. I could have easily proved my point by flying under just one bridge, and that would have been the end of it — but that wasn't my style. Instead, I had brashly committed myself to flying under all four, a feat never before attempted in a landplane.

At seventeen I was the youngest licensed pilot on Roosevelt Field and had been, since my solo flight at fifteen, the youngest "girl" pilot in the world. It had long since come to my attention that people of accomplishment rarely sat back and let things happen to them. They went out and happened to *things*. Ergo, I decided to rattle the slats of the aviation firmament by becoming either the first professional female pilot in the world or the best. Modest little flower that I was, there was never any doubt in my mind that either or both of these titles would be mine. It was just a question of when.

My decision to make today's grand slam was reached when it was pointed out to me that flying under one bridge successfully *could* be looked upon by some as a happy accident. Taking on all four demanded skill and *chutzpah*. Having an overabundance of the latter, I planned carefully to exhibit the former.

The only thing I left to chance was a dry run of the flight itself. In this I had no choice, for the performance was forbidden by everybody — the City of New York, the United States Navy, Coast Guard, Marines, and probably, the Knights of Columbus, the American Legion, and the DAR. Now that the moment of truth had arrived, I faced the possibility that "the best laid schemes o' mice and men/Gang aft a-gley." I also suspected that the Department of Commerce, which hadn't been too eager to issue my license in the first place, might decide to take it back — permanently.

After flying upriver, I dropped the nose, pulled back on the throttle, and positioned the Waco for its descent under the right span of the Queensboro Bridge connecting Welfare Island to Manhattan. During these maneuvers my thoughts drifted back to Herb McCory, the barnstormer, and the bet that had got me here in the first place.

Herbert McCory was the chief aerial photographer for the *New York Daily News,* one of the most prestigious daily newspapers. Mac was one of the few photographers with a nose for news, along with an ability to produce artistic aerial pictures. This talent was backed by a fearless determination to get the best and clearest shots possible — a trait that caused some pilots to decline the chance to fly with him, for when Mac was after a picture, danger was a word he didn't know how to spell.

The whole business about the bridge got started in late September 1928. An itinerant barnstormer from the Middle West had been idling along on his way to Curtiss Field when he spotted the Hell Gate Bridge, a structure spanning the dangerous currents of the Hell Gate Gut just north of Manhattan. Impulsively deciding that this was the perfect way to get publicity heralding his arrival in the East, he shoved the nose of his long-suffering JN-4 biplane down in as much of a dive as she could muster. He had some vague idea of gaining enough speed to zoom out from under the span and loop the bridge for a spectacular finish.

Any East Coast flier could have shot this plan full of holes in five seconds. Sophisticated airplane buffs were bored with bridge stunters, and newspapers were highly critical of such irresponsible actions near public installations. Too many fatalities had occurred from these shenanigans as even the most skillful aerial circus fliers found they could not overcome trick wind currents, engine failures in inverted positions, and other unpredictable disasters. On top of that, a well-known flier named Bert Acosta had already successfully flown under the Hell Gate, so it wouldn't even have been a first — assuming the barnstormer had been able to do it.

After hooking a wing on a stanchion, both the barnstormer and the Jenny splashed down in the swift currents of the Hell Gate Gut. The flier was fished out by a passing garbage scow, while the Jenny had to be raised from the bottom with grappling hooks — a scene which was photographed by Herb McCory and run in the *Daily News*.

Far from being daunted by his dismal performance, our hero's ego triumphed over all obstacles, including the *Daily News*. For what Mac intended to depict as a none-too-bright barnstormer, gazing remorsefully at his wrecked, sodden aircraft, emerged on film as a valiant mission gone awry, despite the lion-hearted efforts of a heretofore unknown but brilliant flying ace.

This farce was compounded when the *News* rewrite man was taken in by the pilot's histrionic power and wrote a sympathetic story that was promptly yanked after a single edition. It wasn't picked up by any other paper, but that didn't prevent Mac's getting an unmerciful ribbing by the fliers on Curtiss Field. Mac's views on irresponsible publicity seekers were widely known and respected, but even the most competent pro can slip up on occasion. Mac ruefully admitted, "He sure put one over on me. But how did I know that the lens hog was going to turn into Douglas Fairbanks at the click of a shutter?"

Mac was telling me the story when a shout went up on the field that Lindbergh was coming in. Mac ran for his camera, and the rest of us took off to protect our planes on the flight line, for the excited crowd could be counted on to swarm over everything in sight. Cracked wing ribs, torn linen, even overturned aircraft were the not uncommon results of a visit by America's hero. From the time of his landing at Le Bourget Airport in Paris in May 1927, Charles Lindbergh was far and away the most famous man of his time. He had become increasingly wary of the press, but Mac was one of the few newsmen to whom he continued to be gracious.

Young, courageous, talented, and handsome, Charles Lindbergh epitomized every girl's romantic dream. Like every other female over the age of three, I had a crush on him. But I had something going for me that none of the others had. Having already met him before he left for Paris, I knew he was so shy of females and so tunnel-visioned about airplanes that his interest might possibly be piqued by my playing down being a girl and confining my conversation to struts, landing gears, and fuselages. Unfortunately, during two brief meetings after his return from Paris I had had no opportunity to display my expertise. Today didn't hold out much promise either. A quick glance at the teeming horde surrounding him, waving autograph books and pens, confirmed that the very best I could hope for would be a brief wave over their heads. I didn't even get that. He either didn't see or chose to ignore my timid salute. Hmmm . . . Lucky Lindy, my eye . . .

one of these days . . .

Mac's problems with the would-be bridge looper were far from over. The Department of Commerce inspector, a feisty gent, was so antagonized by the barnstormer's arrogance that he impounded what was left of the Jenny and suspended the pilot's license, after delivering a stern warning that he was to stay away from airplane controls and confine his future travel plans to bicycle. Far from being chastened, the barnstormer took it all as a personal affront, whining that the inspector "had it in for me 'cause I come from out of town." When he realized he was without airplane, license, or employment, he clung to Mac like the smell of frying.

Short and slight, with watery blue eyes and a straggly, unkempt walrusy mustache, the barnstormer was a pathetic figure in all eyes but his own. Swaggering around the hangar in his filthy water-streaked jacket, helmet cocked over one eye, he drove hapless listeners out of the working bays faster than the noon lunch whistle.

In desperation Mac begged ground jobs for him, but he quickly established that his ineptitude in flying extended to a variety of other areas. He managed to get himself promptly fired from every post Mac got him. Clearly he considered himself above such mundane tasks as pumping gas and changing propellers. He much preferred spending his time perched on the fence at the end of the landing strip, passing out critical comments to airport hangers-on as we pilots came in to land. He had found his niche with these kibitzers. They couldn't fly either.

We had long since tuned out his torrent of verbal trivia, but the day he started grumbling for the thousandth time about how much better he was at the controls than any of us, how if "that ol' OX-5 hadn't started spittin' and sputtin' just as I went under . . ." Mac could take no more. Turning on him with a roar of rage, he bellowed, "It's been done before by an expert, for chrissakes. *When* are you gonna knock it off? Why, even Ellie here could do it, couldn't you, Sis?"

Sympathizing with Mac's distress, I shrugged. "Sure, Mac. Anytime." I was well aware that Mac had directed his comments to me in order to give this male chauvinist the ultimate put-down. His caustic comments about "that kid with the freckles who they let fly around every day while the department cracks down on an old-timer like me . . ." had finally reached Mac's ears. As an old friend of my family's, Mac finally let his resentment boil over.

Neither of us was surprised when the barn-stormer slunk wordlessly away.

I considered the matter closed. But a few days later I was startled to learn that a betting pool had been formed and that I was right in the middle of it. It seemed that this character had gone straight from the Waco hangar over to the Curtiss group on the other side of the field, where he told everyone that I had agreed to duplicate his flight but then had turned yellow and was trying to back out of it.

Part of Roosevelt Field's charm lay in its lack of boredom. Charles Lindbergh's Atlantic flight and the successful flights of Chamberlin, Byrd, and Haldeman that followed it in quick succession gave the field an aura of glamour and excitement that was absolute catnip to the public. It wasn't long before we pilots were also swept up in the pervading atmosphere of eager anticipation, as each day brought announce-ments of record-breaking flights planned, canceled, or in the making. Any place on the map that hadn't been flown to was about to be, and every hangar swarmed with the activities of preparation. We had no notion that we were watching history in the making, for all this uproar drew every promoter, con man, drifter, and fast-buck artist in the area like metal shav-ings to a magnet.

It also drew nuts like the barnstormer, whose unerring instinct for trouble led him to stir up a brouhaha with someone who wanted only to fly her Waco around in the comparative peace and quiet of an OX-5 engine's prop wash.

I had rarely missed a day's flying practice since my solo some eighteen months before. This monotonous toil was aimed at perfecting my skills in order to be accepted by the flying fraternity as a competent pilot. Picking up the gauntlet on this harebrained scheme would do nothing to enhance my image, but those cracks about my flying and being yellow stung me. I took the bait.

A couple of days later, when things had heated up to the "I'll betcha she won't . . . she will, too!" stage, I talked it out with Mac. There was only one way to put this clown in his place once and for all. I would fly under not one, but all *four*, of the East River bridges. While checking the books for tides and bridge clearances, I found that it had never been done in a landplane, so it would be an unusual first. Since I would have to fly the river's length at low altitude, I decided that a Sunday afternoon would be the best time. There would be less water traffic than on a weekday.

I soaked up all the "hangar flying" advice I could find. In the air I religiously practiced low-level figure eights, mostly around ship masts in Manhasset Bay. All of this, plus Uncle Harry's advice on one of his infrequent visits to his sister (my mother), gave me a rough idea of what I was in for.

Uncle Harry was a U.S. Navy pilot assigned to the amphibious aircraft supposedly protecting our shores from foreign invaders. He flew these lumbering, oversized, underpowered behemoths under bridges every day of the week, for the most practical of reasons – he couldn't get over them. According to Uncle Harry, the only way he could have repelled any invaders was to land on top of them, and low-level flying was a good way to develop the technique. These activities made him a mine of useful information, and I learned from him all about low-level turbulence, actual airspeeds versus ground speeds, air cushions close to water, and computing true horizontal clearances. This last information was especially important because the clearances provided by the Coast Guard neglected to mention the fact that bridge spans curve downward, lessening the horizontal clearance.

As the fateful Sunday neared, my fears of losing my license loomed large. That there would be a disciplinary hearing I had no doubt. Much depended on the attitude of the Department of

Commerce and the newspapers. As I now saw it, the only way I'd hang onto that treasured piece of cardboard was if they gave me straight As for the care I'd taken in the planning.

First – the airplane. The Waco I planned to fly was smaller than the Jenny flown by the barnstormer and far more maneuverable. Her thirty-foot wingspan as against the Jenny's forty-three feet greatly increased my chances of getting around (or between) any surface vessels.

Second – making the flight on a Sunday would be my answer to any charge of irresponsible endangerment to life and limb of those on the river.

But the moment of truth would be the flight itself. I'd tried to cover all the bases in my practice sessions, but what if – heaven forfend – an unexpected cross wind slammed me into the bridge abutment? Well, I'd have to give that one my undivided attention at the proper time. Anyway, I'd be wearing my lucky sneakers.

In spite of all I could do to hold it back, Sunday dawned bright, windy, and clear. I donned my brother's old linen knickers and a pair of his argyles, plus one of his rattiest summer shirts. This getup was topped with my oldest, most faded red leather jacket (all the better to spot you in, my dear, if you wind up swimming the East River) and a tattered cotton helmet that no

35

longer sported loops to hold my goggles in place.

When I got to the field, the Waco hangar was deserted, but my ship was out on the flight line, her engine idling. Father's flying instructor, Red Devereaux, climbed out as I approached and stalked off, muttering darkly about "calling off this damn fool idea." Grumpy as he was, I wished he'd stick around for a minute. I had the flight line all to myself. I revved up the OX-5 to clear out its sinuses and mentally reviewed the troops.

The Curtiss crowd was warily eyeing me from the doors of their hangar. That didn't bother me too much. They were betting against me anyway. But where were Mac and all the good guys, who were supposedly betting *on* me? I didn't expect a bottle of champagne to be busted on the Waco's nose, but I hadn't anticipated this kind of freeze-out either.

Obviously I would do battle on my own. I wasn't about to cry — that could only fog up my goggles — but I can't deny that I was feeling just a little sorry for myself when a shadow fell across the cockpit. Someone was pounding me on the shoulder and shaking my hand. I found myself staring into the handsome face of the world's hero, Charles Lindbergh. He was grinning warmly and saying something about keeping my nose down on the turns. I couldn't manage to do more than just stutter my thanks before he loped off, long legs galloping ahead of the gathering spectators. The thoughtfulness of his gesture was so heartwarming that the Waco and I soared aloft like a couple of dry leaves in a high wind.

I flew southwest of Garden City to Brooklyn because I was planning to fly up the river. Circling the Lower Bay, however, I decided it would be unwise to take on both the Brooklyn and Manhattan bridges at practically the same moment. I hadn't fully appreciated how close together they were. There was a much greater distance between the Williamsburg and Queensboro bridges, so I decided to fly over the bridges and come back downriver heading south.

Just north of Welfare Island (known today as Roosevelt Island), I banked around steeply and started under the right span of the Queensboro Bridge, which connects the island to Manhattan. This span has a clearance of 900 feet, and the last tidal report from the Coast Guard indicated my vertical air space at center span was 130 feet. Talk about shooting fish in a barrel!

Peering under the Waco's radiator (awkwardly positioned under the top wing), I was dismayed to find a yacht heading for the air corridor I'd staked out for myself. Slamming the throttle open and yanking the stick back in a single movement, I made the Waco leap the north tip of the island like a gazelle heading for cover after spotting a safari.

After quickly diving down, I hunkered close to the river's surface on the side of the bridge that connects it to Long Island, settled into the smoothness of Uncle Harry's air cushion, and started through the 750-foot span. When I glanced up, my heart stopped, dropped into my stomach, and sank through the bottom of the cockpit. Dead ahead, a gaggle of cables with large wooden blocks attached hung low under the span's curve.

Keeping my eyes glued to the cables, I dropped down to within ten feet of the water and carefully threaded my way past. Popping out of the shadows into the bright sunlight, I realized that my lungs were bursting and my toes ached with cramp. In those tense seconds I'd been holding my breath and riding the rudder bar like a ballerina on point. Gratefully sucking in a lungful of air and wiggling my toes, I was momentarily relieved until I realized the meaning of a wildly waving white scarf on the bridge that I'd glimpsed in the split second before going under. There was no mistaking the wide grin and gingery forelock of its owner — Dave

Oliver, majordomo of Paramount Newsreels. Wherever Dave was, that newsreel truck was right behind him.

How would I ever explain that to the Department of Commerce inspector? It could be interpreted only as flaming evidence of a carefully planned publicity stunt. But right now, with the Williamsburg Bridge looming up, I had to concentrate on Uncle Harry's instructions.

"Stay just above stalling speed. You'll have better control and maneuverability around boats or barges. If you try it at top speed, you can't correct accurately for turbulence or obstructions. Remember, the airplane can do only what you direct it to."

Excellent advice, and it worked like a charm, despite its being counter to everything I'd ever been taught. Even Russ Holderman, my solo instructor, sternly admonished that when I was faced with low and close flying, "Slap that throttle through the fire wall and get the hell out of there!" Of course, Russ never pictured me flying under any bridges. I opted to stay with Uncle Harry.

The Williamsburg Bridge has the greatest horizontal clearance of all – 1,536 feet. Throttling back, I approached it in leisurely fashion. The fat was in the fire now anyway, so I rocked the little ship from side to side in salute to the Fox Movietone crew I clearly saw focusing their cameras on me.

Two down and two to go!

The Manhattan Bridge came next, and I could make out still another newsreel truck — probably Pathé. Behind it were some wildly gesturing spectators. The word had gotten around.

Once more I waggled the Waco's wings, this time adding an arm wave in salute. With the Brooklyn Bridge directly in back of the Manhattan, I was practically home free. Imagine that character back on the field telling everybody I'd never make it. This was going to be a piece of cake.

And it almost was, until the one thing I feared suddenly came to pass. The tanker on my right plodding out to the open sea posed no problem. A deepwater channel kept her to the Staten Island side of the bridge arch, while I was boring through the middle. But without warning, a navy destroyer hove into sight, heading north through a channel that cut smartly into my air space. There should have been ample room for all three of us, but with victory so close I was taking no chances.

Upending the Waco into a vertical bank, I rode the rudder and stick with all my strength and completed the trip sideways, sailing straight over the heads of the open-mouthed white hats grouped on the ship's fantail.

Bringing the controls back into neutral position, I reversed the rudder and slammed the throttle open so hard the little knob almost flew off. Roaring down New York's Lower Bay, I headed straight for the Statue of Liberty. She didn't bat an eyelash as I flew past her nose and pulled up into a steep climbing turn. Looking back, I could see jets of steam rising as the boats in the harbor blew their whistles in salute. I dove down, circled the Lady of the Harbor once more, and streaked for home.

I was grateful for some uninterrupted time to think before landing back at the field. Mac was going to have some explaining to do about the presence of those newsreel trucks. Now that McCory's boys had won their bet, I would have to muster everyone's support in hanging onto Private Pilot's License No. 3178. I looked down at the hand that had shaken Lindbergh's and decided not to wash it for at least a week.

Up ahead, the field came into view. I carefully started down into the flight path, side-slipping in over the telephone wires for a short landing in back of the flight line. This one had to be A-okay, and if I say so myself, it was one of my better efforts.

The minute the prop stopped turning, I was

surrounded. Father and Mac lifted me out of the cockpit amid frantic hugs, kisses, and back-slaps. I had no wish to be photographed in the tacky outfit I'd worn, but Mac was deaf to my pleas, and he and his cohorts clicked wildly away. Any illusions we might have had about our barnstormer's being a sporting loser were instantly dispelled when he ran true to form and complimented me on my luck in getting back safely.

Luck! The nerve of that amateurish bungler, insinuating that the flight had been nothing more than a happy accident. I was furious — but my sense of humor got in the way.

Pretending to polish my fingernails on my sleeve, I blew on them with affected boredom and casually asked, "Care to see it done again — upside down, sideways, or in the other direction?" His stammered reply was lost in the hoots and catcalls from the group behind him.

When I confronted Mac later about the presence of newsreel trucks on the bridges, he happily assumed the blame, explaining, "On Friday we found out this guy was planning to call foul to get out of paying off on the bets. He spread a story around that he'd found out you planned to fly over to Cross Bay and change places with Red Devereaux. Red would take your place, fly over to New York, go under the

the bridges, and fly back here. The only way we could prove you were at the controls was to station the newsreel trucks up there and position them so they could shoot straight down into the cockpit."

I stared at him incredulously. This aerial romp was certainly separating the men from the boys — and the friends from the enemies. According to Mac, everyone stayed away from me before I left so I wouldn't become unnerved at the last minute. Still, it was an eyeopener to learn about a few "friends" who'd tried to pick up a few dollars betting against me.

At home in Freeport that night, the phone never stopped ringing. The newsreels were already in the Broadway theaters. Somehow the public had become intrigued with the flight, and an incredible pitch of excitement was building up. On Monday the newspapers were pretty hysterical, too. As for publicity, short of setting my hair on fire in the middle of Times Square at high noon, I don't see how I could have gotten more.

Despite my relief at the favorable reaction of the public and press, I was worried sick over the possible action of the Department of Commerce. By Tuesday I was a nervous wreck. Worse yet, I seemed to be the only one concerned. Red, Mac, and Father, after seeing the newsreels, were so carried away by my performance that they refused to look at the dark side. "Why should they?" I asked myself bitterly. "The Department of Commerce can't ground them."

On Wednesday the ax fell. Only it wasn't the ax I'd been expecting. A telegram summoned me to a hearing in the office of the mayor of New York City, the Honorable James J. Walker. I stared at it blankly. The City of New York . . . that blows it. My flying career (if I'd ever had one) would soon be over, for the department would undoubtedly be swayed by the mayor.

For the rest of that week, color me blue. I stayed away from the field and took out my frustrations on the piano. Our long-suffering neighbors closed their windows as the crashing chords of Rachmaninoff's C Sharp Minor Prelude assaulted their ears.

Promptly at 10:00 A.M. on the appointed day Father and I presented ourselves at City Hall in downtown New York. We were greeted by Major William Deegan, a handsome graying Irishman who was the mayor's administrative assistant. He chatted cheerfully with us, carefully avoiding the purpose of our visit. His studied nonchalance made my stomach tighten into a hard knot as my mouth turned to cotton.

I had just reached the decision to turn myself

in for a life sentence at hard labor when a dapper, slender little man came into the room. The charm of James J. Walker was legendary even in 1928, but I, at seventeen, was totally unprepared for it. His very presence lit up that gloomy cavern like a torch in a mine shaft.

The sun came out, birds twittered in the background, and my despair turned to hope. Unaware of my mental turmoil, he extended his hand to Father, saying softly, "Hello there. Tom Smith, isn't it? Saw you at the Palace last week. Great show. Enjoyed you immensely. I'm an old songwriter myself, you know. Whenever you need some new material... Tell me, Why are you here? Can I help you in any way?"

The major shot to his feet. "Well, Your Honor, Mr. Smith is here with his daughter. You remember — we sent for her."

The mayor stared at him, eyebrows upraised. "*We* sent for her? Why?"

"Well, Mr. Mayor, she flew under all the East River bridges last week."

Up to now the mayor's attention had been directed at Father. I pulled myself up to my full five feet three inches in an attempt to give dignity to my introduction. I failed miserably. His double take was straight out of Laurel and Hardy.

Turning back to Major Deegan, he said

sternly, "What kind of joke *is* this? You can't mean that this child is the careless devil we are supposed to chastise publicly?"

"I'm afraid so. She's just seventeen, and while it may have been just a prank, we have to do something. Can't encourage people to fly around under those bridges willy-nilly."

"Oh, come now, Bill," chided the mayor. "No one's ever done it before. I'm sure there's a reason. I'd like to speak to Miss Smith alone, if I may."

With a courtly bow he ushered me into his office. The soul of concern, he asked many penetrating questions. I was happy to learn he was familiar with the details of the barn-stormer's Hell Gate fiasco. He also knew Herb McCory.

Under his sympathetic prodding, I told him how I'd begged Father to let me go through with the flight and how he'd agreed against his better judgment, privately telling Mother that he didn't want to put me down in the eyes of the Curtiss group by refusing to let me finish the job. And then the worry over losing my license spilled out in a torrent of words I could no longer suppress.

He looked pensive. "You know, of course, that Bill's right. We do have to take *some* action here."

I nodded grimly.

"Still, I admire your father for giving you the freedom to learn to fly and make your own decisions, even if this wasn't a particularly wise one. I may want to speak to Mr. McCory later, but for the time being, we will issue a statement saying that you have been grounded by the City of New York for a ten-day period. Agreed?"

I nodded my head. My vocal cords wouldn't work.

"This will be done, provided you resist the temptation ever to do anything like this again."

I shook my head vigorously.

"As to the Department of Commerce, you leave Washington to me. I think you can safely count on an off-the-cuff unofficial reprimand and – oh, yes – those ten days will be retroactive from today."

Still speechless, I wanted to dance around the room. Retroactive! I could be back in the air by Thursday – day after tomorrow! Mayor Walker, you are a beautiful human being. Happily my voice refused to return, or I might have spoiled everything.

Then the mayor called Father in, and they chatted and laughed and told stories for the next half hour while the mayor's next appointments fumed helplessly in the outer office.

His prediction regarding the Department of Commerce was right on target. I got the letter of reprimand with a little personal note inserted by the chief inspector's secretary, asking for my autograph by return mail.

2

Planting My Feet in the Clouds

The day after the bridge flight I walked onto Roosevelt Field a celebrity, my treasured anonymity gone forever. But signing autographs and posing for pictures *were* exciting, and when they palled, the press was always there to ask for my opinions or predictions about everything under the sun. How flying a Waco under New York's East River bridges could drape a mantle of wisdom over my youthful shoulders was the one question they didn't ask. Simply put, the temporary crowning of stars was, and is, the way of the American press.

Before Lindbergh flew the Atlantic, they had held his youthful ambitions up to ridicule, dubbing him the Flying Fool. But if President Coolidge hadn't sent the *USS Memphis* to bring him home after the flight, their hysterical adulation would have had the public believing that he had walked back atop the waves.

Through it all, Lindbergh kept his head. While there was certainly no comparison between my initial splash in the papers and his continuing glory, I found a hidden common denominator. We both had sensible mothers.

An unusually euphoric article about me appeared one day in the *New York Journal-American*. I was presented as the new leader of my generation – a cross between Carrie Chapman Catt and Joan of Arc. From then on Mother forbade any more interviews – at least for the time being.

At first I think she enjoyed all the clamor over my flight since the emphasis was on the feat itself and my skill in bringing it off. But when opinion makers such as Arthur Brisbane and Herbert Bayard Swope squared off on my youthful peccadillo (Brisbane for, Swope against), she knew I was too immature to understand what was really behind it all. Pictures and quotations from a freckle-faced seventeen-year-old girl who would cheerfully fly a barn door with wings on it was what, in the twenties, sold newspapers; it was *news*.

But all this attention was heady wine, and I didn't bow quietly to my mother's wishes. I was falling into the fatal error of believing my own publicity. But as usual, Mother laughed me out of it. "Come on now, Sis," she said, "how can I ask Joan of Arc to clear the table or dust the living room?"

Newspaper articles and pictures started to drift in from such unlikely places as Germany, Italy, Australia, and Siam, but Mother reminded me that newspapers, once read, were thrown out or used to wrap fish in, and she also pointed out what I already knew to be true: that one successful flight hardly qualified me as female ace of the year. It wasn't long before I grudgingly went back to practicing figure eights and shooting landings and forgot all about Joan of Arc.

When my parents first summered there in 1912, Freeport, New York, was a sleepy Long Island town. But by 1914 it had become a bustling community. What brought the change? Actors – lots of them. Only thirty-five miles from Times Square, this cool, breeze-swept village so close to the sea was like an unexpected glimpse of heaven for the hotel dwellers starring on Broadway in those pre-air-conditioned times.

Far more accessible by train and auto than the Jersey resorts, it also had friendly natives, something that actors didn't often find back then. After much serious discussion Mother and Father decided to build their home there. They would be close to their friends, and the salt air would be good for me, their two-year-old. Father's career in show business was soaring, and Mother was expecting the next family member early that fall of 1914, which would coincide with the finishing of the new home in Freeport.

Unlike many show business couples, Mother no longer performed. The stage lost a fine singer, but if she had any regrets, they were never mentioned. Father was a talented dancer and comedian. He had been performing the part of the Scarecrow in a touring production of *The Wizard of Oz,* a part being danced in the New York production by Fred Stone, a friend of Father's and one of the finest eccentric and comedy dancers of his time.

Still, Father's own dancing was so electrifying that a Cleveland critic refused to believe that the bundle of rags tossed across the stage by the Cowardly Lion at the end of the first act wasn't just that – not a dancer in Scarecrow costume. A private performance was arranged for the skeptical critic, who then wrote such a glowing review that Father was signed to star in the British production.

The popularity of L. Frank Baum's immortal tale almost assured its theatrical success in London. But Father had never appeared there, and he was nervous about his reception by British audiences. There was also the threat of war. German U-boats were busily patrolling the North Atlantic, and the April 14, 1912 *Titanic* disaster was still fresh in everyone's mind.

Although he very much wanted Mother with him for what they both felt was his big chance, he was reluctant to take her to Europe.

Mother listened to his fears, squeezed his hand, and said she felt perfectly fine about going. Furthermore, I was too young to be left with anyone, and if she were denied this opportunity to see Westminster Abbey, the Tower of London, and Tussaud's, any war would be a Fourth of July sparkler compared to the explosion that would greet him on his return! In late April 1914 the three of us boarded the HMS *Olympic* and sailed for Great Britain.

The London engagement was a hit. The company's run was extended, and negotiations started for a Paris appearance. This, too, was wildly successful, even with no one in the cast speaking French. But once again Father's dancing was the outstanding attraction. Business was standing room only, and it looked as though we would be living in Paris indefinitely when the "shot that echoed round the world" rang out in Sarajevo.

The assassination of Archduke Franz Ferdinand triggered rioting all over Europe. In France an inflamed populace demonstrated wildly in the streets. Young as I was, I still retain a vivid memory of being shoved into a doorway by a young gendarme who shielded Mother and me as a mob tried to break through police lines. The hooves of rearing, frightened horses threatened to come down on their heads, and when shots were fired into the air, panic spread through the crowd. How Father found us in the midst of this chaos is a mystery, but I remember being snatched up in his arms and feeling his hot tears of joy falling on my small face.

We were packed in a matter of hours and on our way to Cherbourg. That night we once more boarded the Cunarder *Olympic,* the last ship to leave France before the formal declaration of World War I. We made the crossing under blackout conditions, with everyone holding his breath, afraid that even a glowing cigarette tip could divulge our position to the U-boats everyone was sure were lurking beneath the sea.

The prolonged tension exploded into a mass demonstration of giddy relief as we entered New York Harbor and steamed past the Statue of Liberty. People wept; a band played. Father was openly emotional. Mother was calm. I waved an American flag and added my childish voice to "The Star-Spangled Banner."

My brother Joe was born shortly after our return from Europe. We moved then from New York City to our new home in Freeport, which should have been finished at least a month before. The contractor soon wished that it had been.

When Mother saw the tangle of wires and planks that were supposed to be a house, she took the builder by surprise. Instead of dissolving into tears and pleading with him to live up to his agreement, she produced the contract in which she had had her attorney insert a penalty clause that the contractor airily assumed she wouldn't know how to enforce. In no time at all an army of painters, plasterers, carpenters, plumbers, and electricians was scurrying around to get the job done.

Scenes like this were to become common in Freeport, for the town's economy soon relied heavily on the building and maintenance of homes for its theater colony. Freeport was at first delighted with the flow of money and talent brought by the influx of performers, producers, and composers, who were in turn appreciative of the town's bucolic pleasures and zephyr-cooled nights. Cooperation was the word of the day, and the theater people readily offered themselves for local charity shows and benefits.

They had first learned of Freeport through the stomach of one man – Diamond Jim Brady. Visits to its dining and dancing pavilions by this gourmet, with glamorous Lillian Russell on his arm, focused his friends' attention on this discovery sometime around the turn of the century.

Close as Freeport was to the big city, getting there in those far-off times was a journey of several hours either on the Long Island Rail Road's stuffy coal-powered trains or by car on the single-lane Merrick Road. Driving, if one drove, there would usually be several flats along the way, for tires punctured easily in those days, and motors continually overheated. It probably took more time to negotiate those thirty-five miles than it does to get out to the Hamptons today, but it was this very difficulty that protected Freeport's charm.

Up to 1900 its summer colony had consisted of wealthy politicians and publishers, whose stately homes, surrounded by manicured lawns and carefully clipped shrubs, occupied the central part of town. Trees tall enough to form overhead arches shaded its streets, and the town's air of peaceful tranquility was disturbed only by the occasional clip-clop of a milkman's horse.

The waterfront area was far more lively. The South Shore Yacht Club boasted tennis courts, and there was a busy riding academy nearby. Boating in pleasure craft from canoes to yawls attracted more than the local boatowners, and during the peak summer months the harbor was jammed with self-appointed admirals. The town's perimeter, laced by a network of canals, provided ample space for private beaches and docks, something that appealed strongly to performers confined for so much of the year to the inside of theaters.

Leo Carrillo and Victor Moore were among the first to own homes there. They were soon followed by Fanny Brice, Sophie Tucker, Arthur Deagon, Fred Stone, Helen Broderick (Broadway comedienne and Broderick Crawford's mother), Moran and Mack (The Two Black Crows), and Will Rogers. At one time the town was literally the Hollywood of the East Coast, filled with actors, writers, producers, agents, and composers. The songs for next season's Broadway shows wafted out on the evening air from the homes of Bert Kalmar, Harry Ruby, Richard Whiting (Margaret's father), and Harry Von Tilzer, all popular composers of the day. Rae Samuels, Belle Baker, Grace Hayes, and Sophie Tucker could be heard practicing their songs for the coming year. Garages and family rooms were pressed into service for the rehearsals of Pat Rooney, Duffy and Sweeney, the Four Mortons, McKay and Ardine, Eva Tanguay, and Trixie Friganza as they worked on lines and timing.

In our home Father's fellow comedians were frequent visitors, and dinner was always followed by an evening filled with stories and good talk by W. C. Fields, Arthur Deagon, Buster Keaton, Bobby Clark, Jimmy Duffy, Tom Dugan, and many others lost to memory. It was the day of clean comedy, and these artists had nothing but scorn for comics who stooped to vulgarity. Body English, facial expressions, and timing were the altars they worshiped at, and woe betide the thief who tried to encroach on their territory or steal their material. Looking at the stand-up comedians of today, one wonders where all that glorious humor has gone.

In the course of these long after-dinner talks, the work of their fellow actors was always a favorite topic. But in general they all agreed that Charles Chaplin towered over everyone. That is, everyone agreed but W. C. Fields. He stopped coming around. Father told me later it was because he was angry that Father hadn't argued that he, Fields, was the greater of the two. Father felt bad about the incident because he had great respect for Fields's work, but in his opinion their talents were too different to allow comparison.

Being allowed to listen to these master craftsmen talk about their work gave me a lifelong appreciation for the timing and skill required in the performance of good comedy. I later learned that similar skills and timing have to be developed to fly airplanes successfully (in the absence of rule books and flight manuals), a fact which gave me even greater respect for those seasoned performers who adapted their skills daily to each new audience with nothing more tangible to lean on than intuition and experience.

The time came when Freeport's performers wanted a club of their own. It had become apparent that the town welcomed their money and exploited their talents but withheld social acceptance. Yet the town fathers turned down the performers' building application again and again, advancing all kinds of reasons why the erection of a clubhouse was impossible. Most of these reasons had to do with the fear that there would be wild parties, scandals, etc.

The performers felt that this was a ploy for an expensive payoff, especially since the town fathers were not entirely untarnished themselves. So they compiled a list of the old guard's wild parties and scandals, which included pictures of a leading banker's daughter at a midnight beach party, where she was clearly seen to be both tipsy and naked. Opposition to the club collapsed after that, and plans went ahead for its construction on a choice piece of waterfront property. The LIGHTS Club, or the Long Island Good-Hearted Thespians' Society, was soon in business, and so it remained to the end of its days many years later.

In all fairness to the town fathers, the performers protected their own interests in ways that can only be termed unique. One of the funniest was the brainchild of a famous film comedienne who later starred in Hal Roach comedies. Protesting an exorbitantly padded garage bill, she drove topless to the beach club where the mother of the offending service station owner was entertaining friends. The bill was immediately readjusted, and our heroine turned her attention to more important matters.

One of the highlights of the LIGHTS Club

was its Christmas party and parade, held in the summer because most of its members had to work the Yuletide holidays. The parade became one of Freeport's annual attractions – the closest thing to a Barnum and Bailey spectacle that talent and ingenuity could make it. Clowns, acrobats, a marching band, and, one year, some borrowed elephants made it an event of national interest.

We all rode or marched in the parade prior to repairing to the clubhouse for gifts, refreshments, and a chat with Santa. The clubhouse had been constructed with a second-floor veranda facing the bay. After trimming the tree and opening our presents, we walked out on it to eat our ice cream and watch the wheeling gulls fight over the cake scraps thrown out onto the water.

Becoming a homeowner aroused all of Father's do-it-yourself instincts. He was convinced he could fix anything, but somehow the inanimate objects he was dealing with always fought back. A simple task like picture hanging became a major project, with Mother, Joe, and me standing by to hold frame, tools, ladder, and hooks while he punched holes in the plaster or mashed his fingers. We never did figure out how he broke his thumb raising the car hood or sprained his ankle repairing the screen door. This last mishap finished his handyman's career. He was forced to cancel three weeks' bookings to the tune of $1,000 a week.

Mother greeted change or challenge with the calm conviction that there was always a solution at hand. Father regarded anything out of the ordinary as a personal vendetta by the gods and charged head-on into the fray, throwing punches and invective with equal abandon.

As individual in appearance as they were in disposition, Mother was a natural blond beauty with gray eyes and a complexion that scorned cosmetics, while Father's jet black hair, blue eyes, and slim athletic figure never betrayed the fact that he was ten years her senior. They were a unique pair, a mutual adoration society that grew stronger with the years and never faltered.

In those days Freeport was a vacationer's dream and a growing child's paradise. The fishing was unsurpassed. A line casually cast from the dock would reel in fluke, flounder, blackfish, snapper, even eels of incomparable flavor. The same line might surprise you with an exotic but inedible blowfish, sea robin, or hackleback. For the hardy souls who ventured out to the deeper waters, horse mackerel (as tuna was then called), bluefish, striped bass, cod, and sea bass were there for the taking.

Clamming was the town's oldest industry. Steamers, little necks, and large hard-shell chowder clams could be bought at the docks for pennies a bushel. Peerless blue-leg and soft-shell crabs thickly populated the canal bottoms, and they, along with sea scallops, shrimp, and lobsters, were among the succulent delights yielded up daily from the sparkling waters of the surrounding bays and inlets.

There were still large tracts of farmland in outlying areas that kept Freeport supplied, via horse-drawn carts, with dewy, fresh-picked corn and green vegetables whose flavors haunt me still. The dairyman made daily deliveries of milk, eggs, butter, honey, and cream so thick you had to spoon it from the bottle. Once a week he brought poultry – chickens, capons, turkeys, and Long Island ducks. Berries of every sort grew thickly in the open fields surrounding our home, and after the first production of Mother's blackberry pie we needed no urging to scurry for the berry pails.

The quality of wonderfully fresh ingredients eventually tempted Mother to take her first tentative steps toward the stove. She was a city girl who had difficulty timing an egg that someone else was boiling. But three or four years after we'd settled in Freeport she decided it was time she learned how to cook. After subsisting on restaurant and hotel food for much of the year, Father went into such ecstasies over her initial efforts that our cook quit in a huff.

This was a potential disaster because Father, who enjoyed playing host, had grandly extended numerous dinner invitations, and now there was no cook! Mother tried to hire another, but the summer season was under way, and none was to be found. So, armed with the new *Fannie Farmer Cookbook* and three helpers — her own mother, whom we called Gram, my brother, Joe, and me — Mother launched the first Operation Dinner Party.

Before we knew what was happening, we were learning to shuck corn; hull peas; make radish roses, cucumber scallops, and celery curls; chop chowder clams; scale fish, skin eels, and crack lobster claws; and even enter the exalted realm of sauces. As the eldest (I was now at the exalted age of six or seven) I was given the privileged job of whipping the egg whites with cream of tartar for Mother's meringues. I loved whipping the heavy cream, too, until the day I got carried away and made a buttery mess of it. From then on Gram timed me carefully.

Our efforts paid off in dishes that were a delight to the eye as well as a poem to the palate.

It wasn't long, though, before Gram began to bend the rules. An excellent cook herself, she resented Mrs. Farmer's stern directives to "level off each spoonful of dry, and measure out by dram each of the wet ingredients." Gram, who measured everything in pinches of this and handfuls of that, had no patience for this nit-picking. Her ire got the best of her while she was making a hard sauce that Fannie directed to flavor with brandy "added by droplet to the taste."

"Humph!" she snorted. "You'd think a body had no sense at all. When I make a hard sauce, it's hard!" Wielding the brandy bottle, she let it gurgle into the sauce as Mother winced.

Fannie wanted the kitchen spotless and ready for military inspection at all times, but Gram viewed the stove more as an altar that required only one priest, whose privacy was inviolate. When Mother once protested about baking her famous twelve-egg raisin-rice custard in our porcelain dishpan, Gram became quite snappish. "You know well enough, Agnes, this recipe can't be cut in half. The dishpan is the only thing in the house big enough to hold it. Who's to know what it was baked in if all of you keep your mouths shut?"

She was right, judging by the groans of pleasure from the guests who later spooned up this creamy masterpiece topped with huge dollops of clotted cream. They wouldn't have cared if she'd baked it in an old crankcase.

By summer's end Mother had stopped yo-yoing between her two mentors. As her own skills developed, tensions in the kitchen disappeared. Now that our cram course in cookery was over, Father wanted to reward us all in some way.

As an avid baseball fan Gram wanted to see the upcoming World Series. Exhausted from her labors, Mother remarked somewhat tartly that Yankee Stadium wasn't for her and that she would much prefer a new living-room rug. Nor did she think that Joe or I were old enough for the nine innings of big-league baseball. The upshot was that Gram and Father had a wonderful time at the Series, Mother bought a lovely rug, and Joe and I got what we *really* wanted: a ride in an airplane.

Father had flown in California in 1916 while he was on tour, and it was in his enthusiastic letters home that Joe and I first got a hint of the joys of life aloft. If Father encountered any hazards in this new mode of travel, he failed to mention them. He wrote, instead, that it was the most glorious adventure he'd ever experienced – and that was enough for us. We couldn't wait to try it.

It was the custom in those days for families with cars to take long Sunday drives and compare notes with their neighbors on returning. Joe and I were engaged in our usual pillow fight in the rear of Father's new all-white Mitchell sedan (a chrome-slathered monster that drank oil like a parched elephant at a water hole) on one of these Sunday drives when Father suddenly pulled up alongside an abandoned potato field with a sign stuck in the middle of it:

AIRPLANE RIDES — $5 & $10

Resting on the grass behind the sign was an odd-looking craft that to Joe and me was as spectacular as something out of *Star Wars* is to children today.

I recently looked at an old drawing of a Farman pusher type of biplane like the one we found that day in the Hicksville potato field. The picture of this bizarre-looking craft forced me again to marvel at our pilot's skill. He was Louis Gaubert, who, I have since learned, was one of France's leading fliers and was then representing the Farman brothers in this country. Gaubert's expert handling of the Farman pusher enhanced his own reputation as well as the aircraft's.

It was a curiously assembled airplane, with a bullet-shaped cockpit jutting out in front of an uncovered but strutted and braced fuselage. The exposure of struts and longerons to the open air gave the whole plane an appearance of fragility that was far from reassuring. If the imposing power plant, propelled by a rear-mounted propeller that looked capable of towing a Cunarder across the Atlantic, had broken loose from its precariously elevated moorings, we would have been promptly decapitated. Happily this ugly thought never crossed our minds.

I remember Father tying my blond braids together to keep them from blowing around and buckling the seat belt over both of us. I don't recall much noise — maybe the pusher engine installation blew it all out behind us — but what I cannot forget is the view. It was totally unobstructed, and its beauty was breathtaking. Shafts of sunlight streamed down through broken clouds, turning the drab truck farms below into a fairyland of gilded greens and golds. Constantly changing cloud patterns were seemingly painted on a vast blue canvas by an unseen artist who directed the wind to airbrush his designs. I wanted never to leave and begged to go up again as soon as we landed. Gaubert looked over at Father and smiled. He gunned the engine, and the ground dropped away once more. The second flght was better than the first. He climbed higher and took us farther. If I weren't already in heaven, it would do for now.

By the end of the summer Joe and I were such enthusiastic shills for Gaubert's passenger-carrying venture that he credited us with turning his whole season around. After all, what could be more reassuring to a hesitant passenger than the sight of two chubby towheads giggling delightedly as they were lifted from the cockpit into their parents' arms?

During one of those early flights Gaubert placed my small hands on the sawed-off wheel that served as a control stick. After letting me "steer" for a few moments, he solemnly informed Father, "She will fly one day, with the great ones. She has the touch." Impressed, Father repeated this pronouncement to Mother on the way home, with all the reverence one might accord an engraving on the Rosetta stone.

Mother murmured, "Mmm . . . That's nice, but how can I get those awful grease spots out of her new pink dress?" He looked at her despairingly and shook his head. Sometimes she just didn't understand.

It is obvious that my love of planes and flying was nurtured by Father's interest, one that he held to the end of his life. He was not an engineer, a scientist, or even a technically oriented amateur. But his emotional barometer was as sensitive as a space-probing radar dish. Despite anyone's pronouncements to the contrary, he knew this was where the future lay and was determined to be part of it.

After his return from Europe in 1914 he was starring on the Keith-Albee Orpheum Circuit. One of its top moneymakers, he was dissatisfied only with the long hours he was forced to spend on trains crisscrossing the United States and Canada, for this vaudeville empire was far-flung. By the time he discovered that the development of aeronautics was lagging far behind the rosy picture painted for the public by its hopeful designers and builders he was as hopelessly ensnared in its dreams of speedy travel as a teenager with a first love. So was I.

That never-to-be-forgotten summer of flying with Gaubert shaped my life. I remember so vividly my first time aloft that I can still hear the wind sing in the wires as we glided down. By the time Gaubert touched the wheels gently to earth I knew my future in airplanes and flying was as inevitable as the freckles on my nose. I also knew that ephemeral as those puffy clouds were, the day would come when I would firmly plant my feet in them.

3

Daredevils and Dreamers

Before Louis Gaubert went back to France, he introduced Father around on Curtiss Field. Father couldn't have been more excited about a personal tour of heaven by Saint Peter. As far as he was concerned, rubbing elbows with aviation greats Bert Acosta, Clarence Chamberlin, Casey Jones, Clyde Pangborn, Roland Rohlfs, and Bill Winston *was* heaven, and now he didn't have to worry about making it to the other one.

Having fought his way up in show business, Father was keenly aware of the importance of versatility and the difficulties encountered in its mastery. His admiration for these men who flew anything with wings and who managed to

71

survive even when the same wings occasionally peeled off was heartfelt to the point of awe, for he knew the hours of monotonous practice and drudgery required to attain their skill.

There was nothing these early fliers didn't turn their hands to, from huge Keystone bombers to featherweight gliders, and there were no tricks of the trade to which they weren't privy. Father was at home here, for the atmosphere was much like the one he knew in the theater. There was no parallel in remuneration, though. He was appalled to find that some of the finest fliers in the world were dependent for their livelihoods on local air show prizes or the selling of passenger rides.

This odd state of affairs existed because the financiers on Wall Street thought the aviation industry was run by daredevils and dreamers, neither of whom were considered good investments. The most venturesome daredevil of all, young Charles Lindbergh, would one day turn their theory inside out and have them pleading for the chance to put up their money, but that day was still years into the future. In 1922 Father also thought the flying business a financial disaster teetering on the brink of collapse — and he wasn't far wrong.

Of all the fliers during the early and mid-twenties, only Bert Acosta ever received a salary commensurate with his talents. When the Curtiss Company hired him as a test pilot, his engineering background was appreciated and paid for. Unfortunately Bert's affinities for the ladies and the grape eventually cost him dearly in a stream of suspensions by the company.

Freddie Lund's situation was much more typical. Freddie worked seasonally for the Gates Flying Circus; that meant for the half of the year that he wasn't with the circus he was "at liberty" to make money by flying passengers, barnstorming, or any number of other high-risk, low-pay enterprises. One of these soon included flying Father around the Orpheum Circuit – at a generous weekly salary with all expenses paid! Freddie was delighted.

"After all," he said with a grin, "why should I be bouncing in and out of cow pastures and ducking Burma Shave signs when I can be living it up with Ethel Barrymore, Will Rogers, and Eddie Foy?" He joined Father in Cincinnati that first winter, and the two of them then flew to San Antonio by way of Louisville, Nashville, Memphis, Shreveport, and Houston. They came back via Dallas, Birmingham, Atlanta, Charleston, Norfolk, Richmond, Washington, and Philadelphia, but they didn't fly out of a single airport on the whole tour. Instead, they had to rely on finding a vacant field near where they wanted to land. Furthermore, the field had to be near a trolley line, so that Father could get into town while Freddie stayed with the plane.

This low-profile pasture hopping was due to an edict issued by E. F. Albee, tyrannical director of the Orpheum Circuit. Father had just hired Freddie to fly him from place to place on the circuit when one of Albee's imperious fiats was posted by the stage manager on the backstage callboard. It read: "All performers making in excess of $750 weekly are hereby notified that to engage in auto or motorboat racing, airplane travel or any other activity considered dangerous by the management or the United Booking Office are liable for instant contract cancellation."

This presumably meant that if you made less than $750 a week, you were free to get killed or injured on your own time. But the headliners whose names drew audiences into the Keith theaters were something else again, and Father was an act in the star billing $1,000-a-week bracket. To put teeth into his directives, Mr. Albee had built up an internal espionage system that could have been a model for Russia's secret police. It was serviced by fellow actors, stagehands, managers, agents — in fact, anyone at all wishing to curry favor with the head man.

Albee was free to banish anyone he chose from working in his theaters, and he wielded his power harshly. The most minute infraction meant an instant summons to his ivory tower over the Palace Theater in New York, where the hapless victim would be summarily sentenced, usually to a long period of enforced idleness.

Since Father had subsidized his good friend and Freeport neighbor Jimmy Duffy (of Duffy and Sweeney fame) for a time when he was out of favor with the czar, he had good reason to pause. Trying to explain to E. F. Albee why he didn't consider flying a hazardous activity was not to be contemplated. Luckily Freddie's expertise and back-of-the-hand knowledge of isolated flying fields were so great that Father never missed a Monday morning rehearsal, and it was a long time before even his faithful pianist, Harry Newman, knew that he was flying, not driving, between engagements.

The Keith-Albee Circuit was the pinnacle of the North American vaudeville world, boasting the Palace Theater (at Forty-seventh and Broadway) as its lodestar. The circuit was unsurpassed in the number and opulence of its theaters as well as in the eminence of its performers. The Palace Theater would remain the Kohinoor diamond in Albee's crown for years to come.

Some of the Palace's old programs reveal that Sarah Bernhardt and Ethel Barrymore played there in one-act adaptations of their dramatic hits. Comedians W. C. Fields, Leon Errol, Ed Wynn, Jack Benny, Frank Fay, Joe Cook, and the Marx Brothers convulsed audiences with their antics. Sophie Tucker, Fanny Brice, Burns & Allen, Arthur Deagon, the Astaires, and Beatrice Lillie sang, danced, and clowned in a display of entertainment talent yet to be surpassed. Fink's Mules, Sharkey the Seal, and May Wirth's Dancing Horses held audiences spellbound with their precise performances, as did riders May Wirth and Poodles Hannaford, star comic equestrian. Long Tack Sam and his beautiful Chinese daughters performed impossible juggling feats. Willie West and McGinty, billed as "The Demon House-wreckers," put on an exhibition of split-second timing barely short of miraculous as they missed buckets of plaster, bricks, and two-by-fours showering down on the stage. And then there was Father. His billing read "Tom Smith, An Artist, A Gentleman and A Scholar," which puzzled the audience until he appeared singing an English ditty and airily swinging a cane. Impeccably dressed in cutaway and pearl gray bowler, he epitomized the British fop. At the end of the song he went into his dance, gracefully sliding the cane into the footlight pan where the resounding crash of the bamboo-painted iron cane broke up the audience. From then on the audience was in the palm of his hand.

Among his more memorable sketches were his "Bumbling Ventriloquist" and "Phony Mind-Reader," both of which were lifted in various forms for musical comedy and; later, for motion pictures. Having written them himself, Father was loath to consider imitation the sincerest form of flattery. He would have preferred the recognition in cash.

Father's reluctance to be drummed out of Albee's distinguished corps of performers in no way dimmed his enthusiasm for flying, as long as he and Freddie could avoid the prying eyes of the old man's spies. Their aerial odyssey continued, and they even managed to get in some flight instruction along the way. Freddie was pleased with Father's progress, although puzzled by his ineptitude near the ground. It was baffling because the moment the throttle was cut and the ground started to come up, Father would shy away and lose all sense of altitude, leaving Freddie — or no one — with the controls.

The reason for this dated back to his first winter in California. He never told anyone that he had taken some flight instruction at that time. Then a student who had already soloed asked Father to become his first passenger. All went well until the landing, when the student suddenly froze on the controls and passed out, leaving Father either to land the plane or to leap over the side. Loath to leave the unconscious pilot, Father tried to make his first landing.

He managed to get down safely by sheer luck and the airplane's innate stability, though he later joked about coming so close to clipping the shingles off a farmhouse that he could read the letters in the mailbox. But it became obvious later that the incident had had a traumatic effect. Freddie worked patiently on the problem at every landing, but by the tour's end he still hadn't succeeded in overcoming Father's ground-shyness.

When Freddie returned to New York, Joe and I begged him to fly us, and then we clamored for stunts, although we pretty quickly wished we hadn't. I don't know what we thought inverted flight would be like, but hanging by our heels, bodies straining against the seat belts as we looked down at a sky spread beneath us, definitely wasn't it.

I objected mostly to the dust, dirt, leaves, loose cotterpins and miscellaneous aeronautical debris that had settled into the tail section and was now flying by my face. Freddie insisted that doing snap rolls was the quickest and easiest way to clean out an airplane and that I could apply this knowledge to my domestic training. By late spring the airplane was internally immaculate and my stomach no longer did figure eights after I got back on the ground. I never told Freddie of my distress. Heaven forbid . . . he might not have taken me up again!

By now summer was coming, and the Gates Flying Circus called. Freddie flew off to join Ivan Gates, and Clyde Pangborn entered our lives. He was a steadier, calmer influence than Freddie, and Father made good progress under his tutelage. So much so that he coaxed Harry Fitzgerald, Father's long-suffering booking agent to keep him in the East — at least until after Christmas. I recall this vividly because it was the first time Father had been home at Christmastime.

All that fall of 1922 my desire to learn all I could about planes and flying was nurtured by "Pang." With infinite patience he answered my questions, attached blocks to the rudder bar so my feet could reach it, and explained the use and function of stick and rudder. Some of it stuck; some didn't. The attention span of a ten-year-old isn't too long, and I was always so anxious to show off what I thought I'd learned that I wasn't too mindful of the next step. He told me later that I talked up such a storm that he didn't see how any of it ever stayed with me. It was only after I learned to listen and stop gabbing that it started to sink in, and then I soaked it up like a sponge.

Words like *fuselage*, *empennage*, *stabilizer*, *fins*, and *ailerons* became as much a part of my vocabulary as *cats*, *dogs*, and *birds*. The mysteries of *dihedral*, *lift*, *drag*, *forward speed*, *stall*, and *spin* had become as uncomplicated as reciting the alphabet and peppered all my conversations.

At the close of each flying day there were "hangar flying" sessions at which the pilots swapped stories about the day's experiences and let off steam. Strangely my presence didn't present a problem, or if it did, those grand gentlemen let me think I was as much a part of the scene as the cracked-up Jenny leaning drunkenly against the back hangar wall.

Bert Acosta was occasionally flying local hops for Father. I was about thirteen when he had me out over the north shore of the Island one day and the radiator of our OX-5 engine suddenly erupted. He said later that I had the depth perception of an eagle and the reflexes of a Yankee outfielder as I unloosed my belt and hung out over the side to spot an open field down into which he could coax the Jenny. Since he was coping with a faceful of scalding steam and I still didn't know how to land one of these contraptions, it seemed the only sensible thing to do. But I treasured the compliment even if such an exaggeration given in the heat of the moment (no pun intended) couldn't be taken seriously. Father almost burst with pride.

It was a marvelous time to be inconspicuous or, failing that, anonymous. There is no better way to learn something than by doing it without pressure, deadlines, tests, or cram courses. In the mid-twenties even the most skillful pilots added to their knowledge almost exclusively through daily experience and shop-talk. The books and magazines then available were written by either scientific theorists or sensationalists. For an adult it must have been a difficult and frustrating time. But for me it was a learning process as gradual and effortless as entering adolescence or developing a bust.

Both my parents firmly believed in opening up as many horizons to a child as she could explore and enjoy. Along with my flying lessons, I was taking French, horseback riding, piano, tap, and ballet. However, they also made it clear that a lapse from the top 10 percent of my high school class could signal a halt to any or all of these lessons.

I would have dropped them all without a whimper except for the flying. They would have had to string me up by the thumbs to keep me away from *that!* Anyway, Freddie, Pang, and now Bert had given me high enough marks to guarantee my stay at the controls — or that's what *I* thought.

My complacency was shattered by a chance remark of Sonny Harris's around the time of my fifteenth birthday in the summer of 1927. Father had just put a deposit on his first airplane, a silver and blue beauty of a Waco 9 biplane that promised to ground the poor tired old Jennies forever. When the papers were signed, there was much good-natured needling about his finally making aviation's "big time," where the private plane owners were distinguished from the renters. Sonny made an on-the-spot application for the job of flying this treasure when it arrived from the factory in Troy, Ohio. (In 1927 a trip from Troy to New York was a long flight.)

Sonny was a huge bear of a man, standing just under six feet and weighing about 250 pounds. His bulk, plus a full shock of jet black hair and beetling brows that framed china blue eyes – their color intensified by a permanently ruddied complexion – gave him an aura of menace that belied both his youth and his personality. He was actually not much more than twenty, and his disposition matched his name. According to Father, it was impossible not to like him and equally impossible to hire him, for Sonny was sowing wild oats at a rate that promised to put the Ukrainian grain crop to shame.

When Father put him off about flying the Waco, Sonny innocently said, "Well, how about putting me in charge of the next one? Ellie's going to solo soon now, and one airplane won't take care of the both of you. . . ."

Father stiffened visibly but then laughed and said he'd think about it. Just then I had one of those intuitive insights I would later learn to trust. I knew he *had* thought about it – a lot. It might even explain why I'd never been permitted to land or take off.

This omission in my training had bothered me for some time. I was now at the point where I could do just about everything else that would be required in straight flying. Unfortunately that can be likened to claiming that you could dance *Swan Lake* at Lincoln Center if only you could stay up on your toes. For takeoffs and landings were, and still are, the bottom line in piloting. The skilled judgment required in those critical moments close to the ground is what separates the men from the boys – or, in my case, from the girls. I had to find out how Father really felt about it. Whining or wheedling never got you anywhere in our house, so it never occurred to me to play "Daddy's girl." Diplomacy was called for, but I was too distraught for that, and as a result, my head-on approach delivered mixed results. The bad news was that there would be no need for a second Waco because I would not be soloing until I reached eighteen. Eighteen . . . those three years stretched ahead of me like thirty!

The good news was that all the men I'd flown with recommended soloing me now with an extended period of advanced flight training to follow. My relief at this disclosure was enough to take the edge off the discovery that I'd been quietly tethered to a short leash by Father's firm instructions to my teachers.

It was obvious that neither he nor Mother understood the depth of my feelings about aeronautics. This wasn't too surprising. I'd never confided my early convictions to anyone. Over the years I'd just happily taken advantage of every opportunity to fly that came my way.

Small wonder they now regarded it in the same light as the music, French, and dancing lessons.

Disappointingly it was Mother more than Father who failed to sense the urgency in my desire to solo. In her eyes the three years would pass quickly, and by the time I'd reached a decision about my life's work she was sure that airplanes would fit in or be cast aside if I'd gone on to something else.

I stared at her as if she'd gone mad. Go on to something else? *What,* for heaven's sake! How could I make her understand that there was never going to *be* anything else? So why wait three more years? Even Acosta had soloed at sixteen – maybe even earlier. Bert was a little vague about it since the first plane he flew he had built himself and to him, chronology was unimportant. He could have been as young as fourteen. He wasn't sure.

Father reasoned: "When Bert soloed, there wasn't any Department of Commerce to regulate things. Its latest figures indicate that eighteen is the average age of solo students across the country."

I failed to see what all this had to do with me. I argued vainly that I *wasn't* average, my training hadn't been average, and my instructors certainly weren't. Furthermore, didn't the last seven years count at all? They listened sympathetically, but the air was heavy with finality. Judgment had been reached, sentence meted out, and the case marked "closed."

Knowing this refusal to let me solo was being done for my own good did little to alleviate my misery, so I sought to spread some of it around. Maybe if I ran away, they'd be sorry for what they'd done and just possibly miss the ray of sunshine they'd done it to.

But then I thought about leaving my newly decorated bedroom, with its pale blue walls, new Circassian walnut furniture, leaded windows, built-in cedar closet, and custom-built dressing table covered in embroidered silk poplin to match the bedspread, with carved sterling silver appointments and a clock gracing its top – and the whole idea lost its allure. Besides, where would I run to – Gram's? She lived in a city apartment. I'd only be crowding her. Sure that she would sympathize with me and take my part, I magnanimously decided it would be extremely unfair to her.

Fortunately it all came to nothing when Gram herself arrived unexpectedly for a visit. As soon as I got her alone, my outrage poured out like hot lava from a volcano.

She looked at me coldly. "Do I understand

you are feeling abused because your parents are trying to protect you?"

Taken aback, I nodded silently. This was my champion? The one person I could rely on to take my side against the world?

"Let me tell you something, Elinor Patricia. You are just about the luckiest girl I know. Maybe that's the trouble. You've been indulged too much. Tom has always been so proud of your progress, telling everyone how you've never caused him or Agnes a moment's worry. I think what you've got is a plain, old-fashioned swelled head. So what if you don't fly that airplane until you're eighteen? Is the world coming to an end? Instead of feeling sorry for yourself, you'd better just thank your lucky stars you've got the parents you have. I don't know of any others who'd put up with you!"

She'd never spoken to me like that before, and it stung me to the point of tears — especially the part about having a swelled head. Maybe it looked like that to her and to my parents, but knowing you're ready for something doesn't necessarily indicate conceit. But what was this inner drive all about? I found that I didn't understand it myself, beyond the conviction that it was something I must do.

Painstakingly I retraced my steps. Even after those first flights with Gaubert, I knew that I would always be attuned to airplanes and flying. Piloting didn't *have* to be a part of that. Had the arrival of that first Waco 9 demonstrator triggered all this? Or were my inner clocks signaling me that it was time to move ahead? Whatever it was, it took brother Joe to deliver the Sunday punch.

"Pop never said you couldn't fly. He's just set against you doing it alone. Didja ever stop to think he may be right? You're still too short to reach the controls without pillows and stuff behind you, and you can't see out of the cockpit straight ahead without standing up. Maybe you better wait till you're taller."

Whoever said, "Some days it doesn't pay to get out of bed," was right as rain, and this, obviously, was one of them.

As long as I could remember, Joe and I had been inseparable. Our natures blended perfectly. I was always the inpatient one, anxious to get on with whatever we were up to. He was the one with logic and reason, who either convinced me that my ideas wouldn't work or beat somebody over the head to give the idea (and me) a chance.

I like to think we would have been firm friends even if we weren't related. Mother said it was like raising twins since she could never discipline just one of us. When I made the balls, he fired them, and vice versa. When we were caught red-handed in some forbidden mischief, bamboo sticks under the fingernails wouldn't have forced us to snitch on each other. So it was not surprising that our flying became a package deal, too. Whatever came out of my present dilemma was sure to affect him, and experience had taught me that he had the cooler head.

As the delivery date for the Waco neared, Father was faced with the necessity of hiring a permanent pilot. He needed a man who was not only an expert flier willing to handle the cross-country trips and a skilled mechanic who could maintain the Waco, but also one who could fill the instructor's shoes vacated by Lund and Pangborn and who would willingly give rides and flying time to Joe and me.

It looked like a tall order until Sonny Harris brought his brother-in-law around. Sonny's relative turned out to be Russell Holderman, one of aviation's solider citizens. He had been prominently identified with aeronautics since his solo in his teens. Now in his thirties, he was looked upon respectfully as the dean of local flight instructors, and his opinions and skill were often sought to settle differences and make recommendations.

After listening to my father's needs, Russ

nodded and said quietly, "Try to get Red Devereaux. He's one of the best students I ever had. I'm sure he can handle all this with one hand tied behind him."

Once more Sonny was the catalyst. He contacted Devereaux, who was now in business for himself, operating a beach-strip aerial sightseeing service near Cross Bay Boulevard in Queens, southwest of the wetlands that would someday become Kennedy Airport.

Edward James Devereaux was a tall, lanky redhead in his early twenties. Because he was conservative, businesslike, and almost gruff in manner, it seemed unlikely that he and my volatile, fun-loving father would hit it off. But hit it off they did, from their very first meeting. Red Devereaux became one of the best friends, staunchest supporters, and most loyal fans my father ever had the good fortune to acquire.

The sobering effect of my two champions' opinions on my solo aspirations lasted until the Waco was delivered. But one look at the graceful little bird, and I went completely around the bend. Each time Red took me up in her, my frustration increased. A few tentative hints in his direction let me know that he was even more set against my hopes than Father. He informed me rather stiffly that he would as soon solo his own sisters, and I gathered that ranked with painting his hair green.

I bided my time until the day I felt I had Mother in a receptive mood. Opening up the subject again was eased by a newspaper picture of two seventeen-year-old girls taken on a California flying field. According to the caption (which I later found was untrue), they had both soloed and gotten their licenses after only five hours of instruction.

I presented my case once more, saying, "See how they'll all be getting ahead of me? And I've got years of flying time under my belt. This is the twentieth century, Mom. Do you want me to start pounding the laundry on a rock?"

She giggled like a schoolgirl and scoffed, "You've never even loaded the washing machine!" But after looking more closely at the proffered picture, she said, "Let me think about it, Sis. It's a big step, and Father will be in Chicago for at least two more weeks —"

"I know," I interrupted eagerly, "but maybe if we surprised him . . . he's had lots of time to change his mind."

"No, I don't think so," she said slowly. "Anyway, you know he doesn't like surprises. . . . But maybe if I call him. . . ."

Elated, I ran out of the house and swung up on my bike. She hadn't said yes, but wonder of wonders, she hadn't said no either. I was careful not to ask any questions for the next few days. My patience was rewarded when she casually brought up the subject again. "If you're still set on soloing, you'll have to report to Russ Holderman. He's got a new field out in Wantagh just off that road they're cutting out to Jones Beach. You have to get there at daybreak to fly over to Roosevelt Field and back. His field is too small to give landing instructions from. Be sure you're back here by seven forty-five so you can get to school on time."

A quick kiss, and she was off to another Garden Club meeting. It had come so fast I was slow to recover. How on earth had she gotten Father to agree? For not only the airplane but also one of the cars would be involved. I'd been driving without a license since I was eleven. Mother had never learned how, so Father had taught me in case there was an emergency while he was away. I had driven her frequently since then to club meetings or the market, but I was always careful to stick to back roads. This was going to mean driving down the Merrick Road every morning when workmen and trucks would be out. I prayed this was an oversight on her part, and as things turned out, it was. She just hadn't thought about that part of it.

As for Father's change of heart, there hadn't

been any. I asked her about this later, and she explained: "There was just no point to calling him. How could I ask him to change his mind? He was absolutely right. On the other hand, I know how I have never gotten over Gram's not taking my singing seriously enough to give me professional training. Up to now I didn't fully appreciate how much flying means to you. Only, Sis" — her voice dropped to a whisper — "do be careful."

It was barely daylight when I arrived at the Wantagh field for the first session, but the propeller was already turning on the parked Waco. By now I knew her cockpit like the inside of my hand, so there was no need for small talk. Beyond showing the signals he would use in tapping his helmet to indicate turns and the more obvious ones for climb, descend, and land, Russ didn't have much to say.

He climbed into the front cockpit, gunned the engine, and then flew straight out over the Atlantic, glistening rosily in the early dawn. He seemed satisfied with my aerial performance, and by the time we were over Roosevelt Field he motioned me to descend. My first landing turned out so well I found myself wondering why everybody made such a big deal out of it. If this was all there was to it . . . I gunned the engine and took off, without waiting to hear from Russ.

When I had climbed to about 1,000 feet, the throttle was suddenly pulled back, and Russ turned around, glaring at me through his goggles. In the next ten seconds he let me know who was boss and who would make the decisions. If I chose to stay aboard, I'd better paste that in my helmet. As for the landing, it was probably a student's fluke, and had I realized my right wing was low on that takeoff? In the future, keep the wings level and your eye on the ball!

It went on like that for the next ten days. I had no way of knowing whether I was doing badly or doing well. Russ's attitude remained noncommittal until I'd had about 3½ hours of actual instruction. (Each individual flight lasted between twenty and thirty minutes.) I thought I was getting the hang of things just fine when he startled me by climbing out of the front cockpit, patting me on the head, and shouting over the motor, "Take her around. She's all yours!"

I stared at him, open-mouthed. More than anything else in the world, I wanted him to get back in there and let me fly him around again. I'd gotten used to that, knowing that if I made a mistake, he was there. I didn't want the responsibility of his cherished Waco on my hands.

94

What if I cracked it up? How would he make his living? What if . . .? A small jeering voice said, "Yeah, how about that? Now that you've got the chance, you're really not sure, or maybe you're just too scared after all."

I put my head down into the cockpit and stared moodily at the tachometer. It really *was* now or never. My bluff, if that's what it was, had been called. Russ had just put his most valuable worldly possession on the line to show me he was confident that I could do it and get it right.

I opened the throttle halfway, kicked the rudder, and taxied back to the edge of the field, giving the engine a sharp burst as I swung the little ship around into the wind. Now pushing it wide open, I watched the tach needle swing to the right on the dial. With the stick forward, the nose went down, and she gathered speed. Before I realized it, the earth had dropped away and I was airborne.

In that instant I knew I was home and would never turn back. Climbing steadily to 1,000 feet, I leveled off and did banking turns to left and right, checking altimeter, horizon, and controls. Perfect! No altitude change, no skid, no falling off, and, most important, no hint of a stall. Better do it again, just to be sure I'd got it all nailed down.

I glanced below and saw Russ's small figure waving frantically. Easing the throttle back, I depressed the nose and glided into a landing. I came in a bit low on the right wheel, but the ship was so slightly tilted that no one but Russ would sense the gentle bump when the left wheel came down. Looking over, I saw him clasp his hands together over his head and motion me around again. We went through this ritual three more times before he came running over to help me out of the rear cockpit and into the front seat. It was one thing to land at Roosevelt Field, but it would be a long time before I was ready to set down cross wind in a little grass patch like his stakeout in Wantagh.

By now we were very late, but I couldn't have cared less. *I had soloed.* Even now, more than fifty years later, I can feel the elation that raced through me as though my heart were pumping pure adrenaline. Maybe it was.

Father's pride in my solo overrode his original objections, although neither Mother nor I had the courage to break the news to him. Russ did the honors, and I never asked for details. The expression on Pop's face the first time I taxied up to the flight line by myself was enough for me.

Luckily I didn't expect recognition for my accomplishment, for any change in my status would escape notice if you so much as blinked your eyes. I might be a female teen-age marvel to the rest of the world, but the Roosevelt Field crowd looked on me as a spoiled kid who had finally gotten her way. They all were from Missouri (literally!), and from here on out I would have to show *them*.

Red continued to be gruffly disapproving, and Russ hammered away at me that all I'd succeeded in doing was open up the can of worms that is the lot of every solo student. For it is only at this point that you begin to understand how much more you have to learn if you are to master the art of flying and survive. Today there is so much information available that you can almost memorize the numbers and get through safely. In the twenties there were no books to crack, so you simply applied the seat of your pants to the cockpit for the start in earnest of a monotonous period of drudgery.

I decided to try to rack up 250 hours of flying time by my sixteenth birthday, which was still three months away. This was the requirement for a transport license, which I couldn't even apply for for another two years. Still, the time would be put to good use. After fighting my way out of cross-wind-induced ground loops a couple of times, I realized fully how much of a lucky happenstance my first landing with Russ really had been. Airplane brakes hadn't yet been invented, so the only weapon we pilots had to counter the ever-present danger of ground looping was a keenly honed sensitivity to air action near the ground. In developing this, I decided to regard the ground as a target and in the next two months got off more shots at it than General Custer with a Gatling gun. During this period the seat in the Waco was permanently anchored forward two notches so my feet wouldn't slip off the rudder bar.

It would be gratifying to report that all this frenzied activity came under the heading of serious dedication, but that wouldn't be entirely accurate. Obsession was more like it. Now that I could fly around by myself, I seemed driven to make up for lost time. Russ was infinitely patient – correcting, explaining, teaching, cajoling, arguing, even threatening, when I became too mulish, with the one weapon that never failed: grounding. He never flew with me during these sessions; he simply held up a mirror to my mistakes. But it was a time-consuming labor and one I will always be grateful to him for.

Red was impressed when I told him that behind it all was my determination to fight my way out of the "student" category. For I knew at last what I wanted. There had been no stunning revelation from on high, no thunderclap followed by a vision, but one afternoon, while idly circling the field, I knew better than I knew my name that not being just another flier, but becoming a *professional* pilot was for me the most desirable goal in the world, and I was not going to allow age or sex to bar me from it.

The tools of this trade, once I'd mastered the necessary flying skills, included such peripheral areas as meteorology and navigation. But when you got right down to it, proven facts were as difficult to seize from the welter of misinformation then available as mercury escaping from a broken thermometer.

Navigation was largely limited to how high you could climb and how far you could see. Celestial navigation was useless to a lone pilot. There was no way to stand up and take "sights" in an open cockpit. There were some self-styled geniuses who claimed to fly compass courses accurately, but the rank and file still did it by landmarks and railroad tracks.

Meterology was catch-as-catch-can. Basically you knew that an east wind on Long Island blew in rain, fog, and the Atlantic's dirtiest weather. The prevailing winds heralding fair and balmy conditions came out of the southwest. Northwest breezes ushered in clear, cold weather, while northeast gales, compounded with snow, sleet, and ice, worsened what the east winds had already dumped on you.

The only place these subjects were dealt with in anything resembling an organized manner was in the military, and even *I* couldn't bluff my way in there. So for the better part of a year I concentrated on honing my piloting skills. Red and Sonny had joined Russ in scrutinizing my landings and air work, and their critiques were of enormous value.

Another of Red's jobs was to help Pop into advanced solo instruction. But Father's solo in 1926 convinced him that he was far from ready to fly completely on his own. Even the fact that his application to it was spotty still didn't explain why his landings by 1928 had advanced only from "poor" to "fair" or, as Sonny succinctly put it, "from disastrous to this side of critical!"

Because he was genuinely fond of Father, Red's concern transcended that of the usual instructor. He had never had a student fail so badly, and he became disheartened as the pile of splintered props and bent wheels mounted behind the hangar.

Professionally Father was going through a difficult time. He had been signed for a feature role in his first Broadway musical, where rehearsals had been troublesome and cast personality clashes frequent. The show was *Golden Dawn* featuring a leading Ziegfeld dancer. It was her first starring production, and many members of the cast hoped it would be her last. This lady could have given Marilyn Monroe lessons in being late, holding up the other cast members' rehearsal times, and in general creating an aura of unpleasantness and insecurity for everyone. Hired to assist in building comedy skits that would fit within the framework of the show's book, Father soon found himself frustrated beyond endurance as his responsibilities to produce sure-fire laughs were increased and his authority to carry this out steadily undermined by the starring dancer and one of the producers. The pleasures of aviation dimmed as he strove to placate these two, and his absences from the field grew more frequent.

During this time Red, probably more to keep himself busy than anything else, concentrated on me. Before either of us was aware of it, I was getting the training he'd been paid to give Father.

I learned more about forced landings than I would ever need to know. He taught me to fly from beaches, roadways, ball parks, and race tracks. Right side up, upside down, cross wind, tailwind, or no wind, I learned how to handle everything from a fire on board to ice build-up on the wings; how to orient myself when the ground was obliterated in fog by finding a thin space and burning a trail in it with my engine heat; what to do if swept out to sea; how to land on water with a landplane and on turf with a seaplane. He omitted nothing and tested me constantly to make sure I retained it.

When I rebelled out of sheer exhaustion, he regarded me coldly. "I thought you said you wanted to fly like a pro. Well, this is how the pros do it. If you meant what you said, let's go!"

There was nothing for it but to snap my goggles down on my nose, slam the throttle open, and follow him out of whatever cow pasture we happened to be in. We flew in separate Wacos, so there was no way I could be relieved at the controls. It was a tough grind, but it seasoned me and forced me to review my priorities. I was scheduled to enter college in September, less than a month away.

Much as I hated to admit it, I needed more time. I could fly every simulated emergency Red dreamed up for me, but I couldn't do it with the locked-in precision that guarantees a flawless performance every time. The skill was on the way. I felt it in my bones, brain, and senses. But as a much-married movie queen said about her weddings, "Dahling, you just keep doing it until you get it right!" I knew it wasn't right – yet.

I argued my case before the family Supreme Court with a fervor that would have put F. Lee Bailey to shame, pleading for a year's sabbatical from academic work. After much discussion they reluctantly agreed, but not before making it clear that my hopes of making flying a livelihood bore little relation to reality. I got the distinct impression that they earnestly wished by the year's end I would reach the same conclusion.

I couldn't wait to get back to the field to share my good news with Red and Sonny. Red rewarded me with one of his rare grins, and Sonny rolled his eyes heavenward in mock despair. "Che-e-e-s! And I thought we were going to be rid of you for a while!" But his bear hug warmed my heart.

As Jimmy Durante would say, "Them was the conditions that prevailed," in September 1928, when Herb McCory came over to the Waco hangar with the scruffy barnstormer trailing a respectful three paces behind, and I ended up flying under the four East River bridges.

4

Blue Skies and Stormy Weather

The Curtiss pilots didn't take kindly to the newsreels and headlines following my flight under the bridges. Overnight I was transformed from "Tom's daughter – the kid with the freckles flying his Waco" into an experienced, publicity-wise competitor. This didn't square with their initial remarks, so when the newsreels showed how I had handled the controls under unprecedented conditions, they turned away from me and made Father their target.

This was cruelly unfair. From the very beginning no one had fought harder than he to keep me out of the public eye so that I could practice in peace. Once he understood the depth of my feelings about airplanes and flying, his backing with ships and instructors had been as complete as money, time, and love could make it. But he never envisioned my flying as a lead-in to a flamboyant career in aviation. For that matter neither did anyone else, including me.

Herb McCory, chief aerial photographer for the *New York Daily News* and a longtime family friend, had been enlisted by Father as early as my solo in 1927 to keep me out of the papers, or, failing that, to hold publicity to a minimum. With public interest stirred to a frenzy by the preparation for the transatlantic flights by Byrd, Lindbergh, and Chamberlin, the story of a fifteen-year-old girl's solo was news indeed. But such premature recognition would expose me to the glare of national publicity, something that Father thought potentially dangerous.

The fledgling stage in flying is the same today as it was fifty years ago, and each student sets his own pace. It is a period of trial and error, of awkwardness advancing (one hopes) toward skill. Each landing is a challenge, and the last thing needed to intrude on one's total concentration is the distraction of the critical scrutiny of a press corps. The easy way out would have been simply to forbid any more solo flying until my sixteenth birthday — more than three months away. But Father said later, "You were doing so very well I didn't have the heart to take it all away from you." With Mac's help, they succeeded in keeping me out of the papers and free of publicity, giving me the time to hone my skills.

Mac had lost a story, but having kept an eye on me from the beginning, he knew that I could no more stay away from airplanes than I could keep my hair from growing, so he gambled that I'd come up later smiling for him and his paper. As a senior member of the press on Curtiss Field for one of the largest metropolitan dailies Mac wielded considerable clout, so Father's wishes were respected. Small wonder that when Mac asked me to oblige him with the bridge flight some eighteen months later, I agreed without turning a hair. No matter how you looked at it, I owed him one.

According to Mac, most of the criticism now being leveled against me could be traced back to Casey Jones, a man both Father and I had regarded as a friend.

"He's just pea green with envy, kid," Mac said. "It's bugging the daylights out of him that you landed on the front pages all over the world. He's also teed off because you're seventeen, you're a girl, you're pretty, and you fly even better than you photograph — all of which he's gonna have a tough time explaining to the Curtiss Company executives. I just found out they wanted to hire you last hear when you broke the light plane altitude record, but he gave them some cock-and-bull story about your being a flaky kid who couldn't be taken seriously because your father is an actor. Come to think of it, that's just about the time that a girl student of his washed out one of their Orioles and never came back. He was probably trying to train her into the job."

It was my first brush with professional jealousy, and I'd be lying if I said it didn't hurt.

That altitude record caused quite a stir among local pilots because of Russ and Red. It all started when someone at a hangar flying session was critical of the Waco's ceiling, saying it didn't live up to factory promises. Rushing to the defense of my graceful bird, I pointed out that 12,000 feet was not exactly a sluggish climb for a biplane of this type and power. When both my mentors got their eyebrows back down and asked how I pulled that figure out of my helmet, I confessed to having gotten up there several weeks back. (I probably hadn't mentioned it at the time because I had no business fooling around up there without oxygen — and I knew it.)

Red was well aware that the stock model Waco had never achieved the ceiling claimed for it, but he still was skeptical. I don't think he doubted my word, but altitude measures were rarely accurate. He and Russ figured it was worth the effort to get an FAI (Féderation Aéronautique Internationale) barograph, and if I could do it again, we'd have a new light plane altitude record. There was much rushing around to get me an FAI license to conform to the rules of international competition, and two weeks later, with the help of a stiff west wind and my lack of weight (at least sixty pounds less than that of any of the Waco test pilots), we had a new light plane world's altitude record — 11,874 feet.

I'd been flying my heart out ever since and wondering why the Curtiss Company didn't invite me to join in the fun when it staged its Sunday flying shows to attract potential passengers. Well, now I knew. I took my troubles to Red, who was his usual sympathetic, endearing self: "Wadja expect from those guys, a medal? Here, hang onto that wrench at the turnbuckle while I tighten up this flying wire. What in God's name were you doing with this airplane, anyway, tying bow knots in the fuselage?"

I was tempted to be truthful and confess I'd been trying to do snap rolls in it when he said,

"You've got to remember, these biplanes were designed as aerial workhorses. Nobody ever said they could do advanced aerobatics." About to reply that nobody ever said they couldn't, I saw that he still had a wrench in his hand and decided against it.

Snap rolls were about as advanced in aerobatics as you could get in a Waco. This sophisticated and violent maneuver consisted of a 360-degree lateral turn done with the nose nailed tightly to the horizon. I was well aware I had no business fooling around with it.

In the twenties Curtiss Field was to the aviation world what the rue de la Paix was to Paris. Every flier of note was bound to turn up there if you hung around long enough. Crowds jammed the field on weekends to catch glimpses of Bert Acosta, Charles Lindbergh, Lloyd Bertaud, Admiral Richard Byrd, Lieutenant James Doolittle, Bernt Balchen, Roscoe Turner, Clyde Pangborn, Clarence Chamberlin, Lieutenant Al Williams, George Haldeman, and Ruth Elder. A place to see and be seen. it was also an excellent showcase for demonstrating skills and exhibiting new airplanes. The famous aircraft designers – Sikorsky, Bellanca, Burnelli, Caproni, Fokker – all sanctioned the display of their latest creations at Curtiss, where the visiting pilot celebrities were often invited to fly them.

The Curtiss Company was quick to capitalize on the notoriety of its fliers. It put together its own air show to attract crowds and sell passenger rides. Staged by Casey Jones, the show was a delightful display of flying talent and expertise. Stunt flying, pylon races, flour-bombing contests, wing walking, and delayed parachute drops all were performed with verve and flair. Designed to appear spontaneous and impromptu, it was as painstakingly choreographed as a formal ballet.

I wouldn't have missed one of these performances for the world, and the Sunday after the bridge incident I was on the flight line, binoculars at the ready, waiting for the show to start, when I was suddenly surrounded by enthusiastic fans begging for autographs. I looked up from signing scraps of paper and girls' diaries to see two boys approaching my Waco with penknives in hand. I had hopped up into the cockpit, with some vague idea of getting away, when Sonny Harris appeared to help me start the engine. Frantically I called his attention to the two boys. Sonny nodded calmly, and the crowd fell back as he made his way to the front of the airplane. Explaining just why carving one's initials into the engine cowling or slicing strips of linen from the wings was forbidden, he got the penknives put away and then lined everyone up in orderly fashion to finish the autographing.

All was going smoothly when Casey Jones materialized, seemingly out of the ground. Ignoring me, he brusquely told Sonny to roll the Waco back in the hangar; we were interfering with the air circus. Sonny looked blank. We were in our usual position on the flight line, nowhere near the point of takeoff for the Curtiss Orioles and Jennies. I had been waiting to leave for Cross Bay to fly passengers with Red and Martin Devereaux when the crowd encircled me, and I certainly objected to having my airplane towed back to the hangar.

By now Red had appeared to find out why I wasn't well on my way to Queens. He quietly asked for an explanation, pointing out that he was in charge of both the airplane and me in Father's absence. Casey mumbled something about our being in the way...air show about to take off...but his voice trailed away before Red's calm authority. He could bluff Sonny, but not Red.

Red turned and waved his arm in an arc to indicate that Sonny should pull the prop around. Sonny and I went through our "contact," "switch-off" routine while Red and Casey continued their conversation. After a few tentative coughs the trusty OX-5 took hold, and I ran it up to full throttle with the chocks still under the wheels. Dropping the motor back to idle, I looked out of the cockpit to see Red's color deepen and his jaw clench as Casey stormed off. Signaling Sonny to pull the chocks, he came over and leaned into the cockpit.

"You *were* practicing snap rolls the other day, weren't you?" he demanded.

Sheepishly, I confessed.

"Feel like doing some today?"

I stared at him through my goggles. Had I done something wrong?

"I want you to get the hell up there and fly those goddamned bastards out of the sky!"

What a relief! Whatever was wrong — and for him to swear like that, plenty had to be — it had nothing to do with me unless, of course, I loused up the snap rolls.

I taxied out and took off. Sonny Harris, not exactly an unbiased witness, reported that the crowd cheered wildly when I finished looping and rolling over their heads, dove down on the field, and pulled up in a flashy climbing turn before disappearing into the southwest. Up in the cockpit I had no way of knowing how my performance had been received. That bit about flying Casey's pilots out of the sky had to be taken with a grain of salt. The day I could fly veterans like Art Caperton, George Pickenpack, and Casey out of the sky was far in the future. Still, I knew I'd done creditably well, and my mood was one of smug satisfaction.

The sandbar airport paralleling Cross Bay Boulevard in Queens was an excellent spot to sell passenger rides, although difficult to fly from. Every landing was cross wind, and the width of the landing strip depended solely on the whim of the tide. Red and Martin had cannily selected this heavily trafficked boulevard as a stakeout for their first business venture. It paid off handsomely from the day they jammed their sign into the sand proclaiming:

PASSENGER RIDES — $5
OVER N.Y. CITY AND THE
STATUE OF LIBERTY — $10 & $15

The rates were as flexible as the rides. The boys made the deals, and I flew as directed. It was a great way for me to build up flying time, even if somewhat illegal around the edges. Still sixteen, I was well over a year away from being old enough to apply for the transport license that would permit me legally to fly passengers for hire. Red chanced using me, saying that, at least in the beginning, he needed three ships

113

in order to make it pay.

As long as he included me, I couldn't have cared less about the details. Nor did it occur to me until later that this was his way of making my flying-for-hire dream a reality. It is incredible to me now that I could have been so dense, but youth has a way of selfishly concentrating on its own concerns, and at that time in my life airplanes even zoomed through my dreams.

As we skittered ahead of the lapping waters of Jamaica Bay down on that beach strip, our Wacos were like those bustling little busybodies of the feathered world, the sandpipers. Red's instructions from the previous summer rang constantly in my ears as I dropped down to hard landings in a few inches of water to elongate my taxi strip or hopped over a sand dune to shorten them.

Flying off dry and wet sand in those days without brakes or oleo strut landing gears presented serious problems at times, but the public was as ignorant of that as they were of being piloted by a teen-age girl — a deception made possible by hair carefully concealed under my helmet, no make-up, and a generous sprinkling of freckles.

I knew that Red would never take any chances with the lives of others. His unspoken vote of confidence in my skill and judgment permitted me to enjoy the work thoroughly. Pilot and passengers always had a wonderful time taking in the sights high above New York City, to say nothing of the beauties of the Lower Bay with its busy traffic in ocean liners, tramp steamers, police and fire boats. The firemen sometimes gave us an added treat as they cleared their hoses by spraying tons of water high into the air, the rising vapor haloed in rainbows of color.

During this period I also got my stripes in commercial cross-country flying. By now Red's duties included overseeing delivery of the planes allotted to the Waco distributor in the area. A request arrived one day for a demonstration of the new No. 10 model for a prospective buyer in Danbury, Connecticut. The only plane available was Father's. Red was busy supervising the assembly of three new planes that were currently in pieces on the hangar floor. One was being cannibalized in order to furnish parts for the other two, so it was a rather delicate operation and one that he couldn't delegate to anyone else. Seeing his worried expression over this conflict in his responsibilities, I was only too happy to volunteer my services.

I can't say he was ecstatic at the prospect.

His grunting acceptance didn't indicate the enthusiasm I would have liked. Still, any kind of acceptance was a step forward. I took off with his admonitions: "Don't get smart, now. No stunts or anything. Just safe, straight flying, and for God's sake, be careful over Long Island Sound. Get up to eight thousand feet as quickly as you can. From there you can *glide* across to Connecticut."

"He should only know," I murmured to myself, "that I've been practicing over the sound and have never been caught yet — and as for wasting time to climb up to eight thousand feet, that would be the day!"

The flight north was uneventful, as was the demonstration flight for the customer. I left for home and almost immediately ran into conditions that had the pigeons walking. The weather closed in on me just as I cleared the town of Danbury. I didn't want to chance getting lost in that prickly country, so I flew east, picked up the Housatonic River, and then carefully wound around each curve at treetop level all the way down to Bridgeport, where I landed to refuel and pick up a weather report from the Coast Guard station.

An hour later they said it was "clearing," so I took off again. Though well aware that our weather reports were often as unreliable as our engines, I also knew that they were the only game in town.

Ghostly cloud fingers stretched landward, giving ominous signs that the ground would soon be obliterated from view. Now heading due south, I decided to drop down to water level but hesitated about crossing Long Island Sound so low. I had opted, instead, to twist my way around every minute bend in the coastline when a capricious breeze unexpectedly lifted the clouds. Ramming the throttle forward, I made a run for it, speeding across the water so low that I could taste the salt spray. Then, when I was so close to the opposite shore that it seemed as if I could reach out and touch the north coast of Long Island, a dark curtain of fog slammed down in front of me.

The north shore of the island is a good bit higher than the middle, where Curtiss Field was. There was nothing to do but pull up into that amorphous mass and pray. Soon a red light winked eerily through the grayness. Was it possible? Could I be lucky enough for that to be the beacon on top of the gas tank at the west end of Mitchel Field? I gambled that it was. Dropping down once more, I found myself skimming low over the farmland in back of the field, an area I knew better than my own room. Nervous perspiration fogged my goggles, but I didn't care. I had made it. I had gotten back in spite of the elements. I was coasting across an approach over the farmlands I could make with my eyes closed.

Flipping my goggles up on my helmet and brushing the water from my eyes, I saw the corner of a hangar glide by my right wing. The tail skid crunched loudly under me as I yanked the stick back into my stomach and slammed the faithful little airplane down to a tail-first landing. Sonny was waiting for me, blue eyes blazing, rain dripping off his nose. Almost yanking me out of the cockpit, he roared, "Where in hell have you been? You should've been back hours ago. We've been worried sick."

When he finally stopped yelling, I explained about the weather and my unexpected Bridgeport stop.

He was unimpressed. "When you pulled up off the north shore, how'd you figure you were flying level? You coulda been spinning, ya know!"

"Sure I could have, but if I was spinning or if I'd flipped over on my back, I'd have had a faceful of cockpit dirt, wouldn't I? Anyway, I had my stopwatch on, and I was only in that stuff for about twenty seconds when I saw the blinker on the tank."

Under such conditions, twenty seconds can

be a lifetime, as Sonny well knew, but so far I'd come up with all the right answers. I fully appreciated the nervousness and concern that made Red and Sonny and even Russ bang and shout to make sure I understood the importance of their teachings. They were trying to keep me alive and flying. If today's performance was any indication, they were slowly but surely succeeding. All the bits of information on my mental tape recorder were being played back in the right order. But I couldn't help looking forward somewhat wistfully to the day when we all could converse again in normal tones.

Red was running toward us with a face like a thundercloud as Sonny intercepted him. "It's OK, Red. She remembered everything. She even checked her watch."

Satisfied that my performance had been no accident, Red rewarded me on the spot, saying that I had earned a place on the roster of Waco pilots qualified to deliver and demonstrate their planes. Just in case there were any male chauvinists in the crowd, I would be listed as "E. Smith, Pilot License No. 3178." The promotion also meant that I would now be paid on the same scale as the others. Who says happiness comes only when the sun is shining?

My instinct was to yell with joy and throw my arms around his neck, but I was still so in awe of him that dignity prevailed. Wet and bedraggled though we were, we solemnly shook hands.

By no stretch of the imagination could this new appointment be called a position. It was strictly a first-come-first-served proposition, depending on a lot of variables.

The area Waco distributor was a crack salesman with the dramatic talent of John Barrymore and the skilled avarice of Willie Sutton. His clients usually came away from the negotiating table with glazed expressions, clutching bills of sale that guaranteed nothing more than future delivery on the latest model Wacos. Not even in the fine print would they discover that these planes were being heavily oversold and that delivery might take upwards of six months.

This was when our piloting skills were called on to placate the wounded, encourage the optimistic, and reward a fortunate few with the actual delivery of a spanking new green and silver Waco 10 biplane. In the beginning I was so low on the totem pole of the piloting roster that I drew only the least important chores, like circling the field in Father's airplane to demonstrate its flight characteristics. In my mind's eye, of course, I already owned the Waco factory and had just dropped in to do the Waco dealer a favor for old times' sake. It was the only way I could keep my spirits up and my ambition down while my air time and skills caught up to even things out.

Like Father, this chap loved flying but had great difficulty mastering it. Short and stocky, with an owlish look heightened by Coke bottle eyeglasses, he looked, according to Sonny, as though he had stepped out of an airplane at 4,000 feet without a parachute and landed standing up. He wasn't around the field on a daily basis because of his profitable sideline, New York City real estate, but he managed nonetheless to fly a good deal. His grandiose promises sometimes demanded his appearance at the Waco factory in Troy, Ohio, but after a few unfortunate experiences with cross-country flying, he was convinced he needed Red along for travel insurance.

With both of them gone, Sonny was in charge. As my flying improved, he followed Red's lead, farming out more and more jobs to me. I had one thing going for me. I was always available. If I wasn't working, I was in the air practicing. When the Waco representative questioned Sonny about using me so much, he replied, "Look, Tom's been one of your best customers. He racked up three Wacos in the past year. But Ellie doesn't fly like him. So far she hasn't even scratched a spinner —

and besides, she works cheap!"

Being entrusted with the delivery of new planes to their owners was heady wine for a beginner. Usually their surprise at finding such a youthful female at the controls was pleasant, but I recall one dour individual in New Haven who insisted on seeing my credentials, claiming he'd been defrauded by not having his ship delivered by an experienced aviator. Displaying my Department of Commerce license did little to calm him, but when I produced my FAI license, enhanced with Orville Wright's signature, his irascibility collapsed like a pricked balloon.

Much as I enjoyed my work and was grateful for the opportunity to plant my toes on the first steps of the ladder toward my goal, caution forced me to realize how shaky these steps were. I had wangled a year's sabbatical from study in order to hone my skills. I still had eleven months of this precious time ahead of me and knew I had to make good use of it. But now came the bitter truth. Even with the summer passenger work and local air shows, I would never be able to make a decent living.

One day I confessed my worries to Red, who in turn spoke to Tex Bohannon, a pilot and delightful raconteur, who had contacts and friends everywhere. In a matter of days I received two moneyed offers for my appearance at the Chicago Air Show, plus a phone call from G. P. Putnam, Amelia Earhart's business manager. He was going to be at the air show, too, and suggested that he, Amelia, and I get together in Chicago.

I was elated. This was more like it! Just thinking about what he had done for Amelia was enough to send my hopes soaring. Amelia had always frankly admitted that her Atlantic flight in the summer of 1928 had been made as a passenger with Wilmer Stultz at the controls. Actually she called herself a "sack of potatoes" on that junket. According to the press releases, she had soloed some years before, but little was known or publicized about the years in between. She had yet to fly publicly by herself, but as a result of Mr. Putnam's skillful handling, she was becoming known as America's Lady Lindy and was reportedly making $500 weekly at receptions and dinners arranged in her honor.

I admired Amelia enormously. Even to be a sack of potatoes on a transatlantic flight in 1928 took gut courage. I didn't aspire to any such adventure. All I wanted was some way that I could hang in there and fly. And I wanted to do it on merit, not because I was a girl, or seventeen years old, or for any of the other reasons the reporters imagined. I wanted to be judged on my piloting skills alone. Red's notation, "E. Smith, Pilot License No. 3178," expressed it exactly.

I looked forward to this meeting on two counts. I was about to meet my personal heroine, and maybe, just maybe, Mr. Putnam would have some ideas on how to solve my problem. Was it too much to hope?

It was, although at first it didn't seem so. With my daily appearances at the show and Amelia's own activities, our meeting was delayed, but we finally arranged to have breakfast one morning in her suite.

My last-minute nervousness was dispelled by her friendly warmth. There was nothing cold-fishy about her handshake either. Taller than I, she was slender as a reed. A tracing of tiny freckles across her nose enhanced a perfectly natural complexion. I hadn't realized until that moment that there was a noticeable spacing between her front teeth.

She giggled like a schoolgirl as she glanced at G.P., saying, "That's why he makes me practice smiling with my lips closed. There's so much to learn about this publicity business." There certainly was, and I was just beginning.

George Palmer Putnam was a handsome man. He was over six feet tall, with snapping dark eyes and close-cropped dark hair, and you could feel his electricity across a room. Wiry and dynamic, he exuded the authority of a bank president or a millionaire, which he was. If that weren't overwhelming enough, his manners were impeccable. Before I knew it, I was agreeing with his implied opinion that our meeting was one of the high points of his long and successful career, and wasn't it a shame that our paths hadn't crossed until now?

With the social amenities taken care of, he got down to business, peppering me with questions as he paced the room. Obviously embarrassed by this brusque change in attitude, Amelia interrupted to say, "Do come over to the table and sit down. You must get a good breakfast into you before G.P. monopolizes the conversation. He never eats much anyway. I think he lives on larks' tongues, and we do have to face those crowds later."

Awed by both of them, I was unable to do justice to the delicious breakfast and settled for buttered toast and tea. Amelia's gracious manner did much to put me at ease, and in minutes we were chattering away like lifelong friends. I was surprised to find that she knew all about me, including my early solo and the FAI altitude record in Father's plane that took so long to be recognized because I was underage at the time. She even knew about my recent work delivering Wacos and laughed sympathetically when I told her the true story of my bridge flying.

Impatient for us to finish, G.P. had long since left the table and was striding up and down in front of the bed. He leaped into a break in our conversation like a member of a relay team changing runners.

"I think it would be helpful for you top pilots to be brought together so that some sort of company could manage your affairs, provide backing for record flights, and afterward arrange for your appearances — at suitable fees, of course."

"You mean something like a literary or theatrical agency?" I asked hopefully.

"No, nothing like that. I'm referring to a strictly professional organization," he said, seeming momentarily angered by my question.

I was puzzled by his anger because he had already talked about contracts for personal appearances, financial sponsoring for record flights, writing, films, and so on — to be handled exactly as he had done for Amelia. But he was reputed to be one of the wealthiest and most influential publishers in the United States, and I thought perhaps people had different terms for such matters at his lofty level. Whatever he chose to call it, it spelled management and professional guidance to me, and I was decidedly disappointed when it turned out that I mustn't get my hopes up.

"This is by no means a definite offer," he cautioned. "I'm sorry if I gave you that impression. We need much more time to evaluate your capabilities. But I can assure you that you'll have our decision quite soon."

"In the meantime," he casually continued, "what are your plans for the future?" He had already let drop that Acosta, Chamberlin, Byrd, and Hawks were among the pilots to benefit from his help. Despite my blighted hopes, it was flattery of the highest order even to be considered for this distinguished company.

"Oh," I protested shyly, "any ideas I might have about my future would be nothing compared to yours."

Months later I told the story of this meeting to Lady Mary Heath, the noted British aviatrix. She doubled up with laughter, shrieking, "Oh, my dear, what I would have given to have *been* there. Just to see his *face*. Dear girl, up to now that approach of his has been foolproof. *All* you Americans are so in awe of this man. You tell him everything, and then he knows exactly how to cut the ground out from under you. You must certainly be his first disappointment."

I didn't believe her until much later. At the time I was too upset at not having passed his "evaluation of my capabilities." His call hadn't come.

But that day in Amelia's suite in Chicago I completely misread the angry or baffled expression on his face. I thought I'd embarrassed him in some way. To relieve my confusion, Amelia started discussing a proposal for the endorsement of a line of sports dresses to be worn by lady fliers in the cockpit. I had to point out that my daily costume was usually an old pair of my brother's knickers, topped with a shirt and windbreaker, legs covered with argyles, feet shod in sneakers. I was scarcely a devotee of fashion, and my outfits leaned toward the practical. I did so much flying over Long Island Sound and the Atlantic Ocean that I wore clothing I could get out of easily in case it was necessary to swim home. I had no experience in keeping dresses and shoes free of the grease and dirt invariably present in the cockpits of the planes I flew, and I wasn't sure I wanted any.

Putnam persisted, "But if you were flying in a closed cabin Fokker or Bellanca, you wouldn't be worried about clothing, would you?"

"Probably not, but I doubt I'll get a chance to find out. To the best of my knowledge, no female has ever been permitted to do more than hold the controls of a Fokker or Bellanca in the air. Certainly no female has ever taken one off or landed it."

He and Amelia exchanged knowing glances. "Oho!" I crowed silently. "I am in the presence of an iron butterfly. But then it takes one to know one."

As my meeting with Amelia went on, I was able to distinguish more clearly the Putnam publicity touch from the very real and warm individual I was facing. The image of a shy and retiring individual thrust against her will into the public eye was a figment of Putnam's lively imagination. Amelia was about as shy as Muhammad Ali. I do not mean to imply that she wasn't modest, as can be seen from the way she characterized her Atlantic flight. But she was already a woman of thirty (something Putnam always skirted carefully in his press releases), who knew definitely what she wanted and where she wanted to go. She left no doubt in my mind that she wanted to go in the same direction as I, but she was wise enough to let me believe that it was going to be ever so much fun to see who got there first.

As I got up to leave, I was still baffled by G. P. Putnam, but Amelia was my kind of people, and I could only hope that I had come across to her as something more than a teen-age marvel who would fade out of sight like the flagpole sitters of the day.

Thinking about out meeting later that night, I kept coming back to the number of times Putnam had subtly attempted to turn the conversation around to my future: "What are your plans after the show? Surely you don't intend to waste your time and talents delivering airplanes? Have you had any offers from the manufacturers at the air show?"

As a matter of fact, I did have one that I was turning over in my mind. It was from the Brunner Winkle Aircraft Corp. It wanted me to launch its new model Bird biplane by setting a solo endurance record in it, just before the opening of the biggest indoor air show yet produced. The show was to take place at the Grand Central Palace in New York City, and one sure way of focusing attention on a new airplane was to come into the exhibition with a newly minted world's record under its belt. Having that record set by the youngest licensed girl pilot in the world wouldn't hurt either. Since I had already put this particular Bird through some of its initial tests, the company could also avoid a further and sometimes costly period of pilot familiarization.

Proving the worth of an airplane by putting

it through its paces in the air was rapidly gaining acceptance. The actual performance of engines under varying conditions of weather and altitude was important to pilots who knew that wind tunnels, while scientifically elegant, could never come close to flight under real conditions.

The one problem with the flight would be the weather. February is the worst of months on the eastern seaboard. I knew I could anticipate anything from a blizzard to a hurricane. I couldn't say I liked the odds, but they weren't insurmountable.

As I weighed the pros and cons of the Bird offer, I found myself, hearing again the rapid-fire voice of G. P. Putnam sketching the plans for his company and the glories of my and Amelia's futures. I now realized that I could no longer drift along, leaving that future to chance. Clearly, if there was any possibility of passing his promised scrutiny, I would have to upgrade the image of a kid flying her pop's Waco under a quartet of bridges. The more I thought about it, a solo endurance flight might be just the answer.

I reached for the phone.

5

Are You Sure This Is How Katherine Stinson Got Started?

Back in New York, I plunged into the rush of activity surrounding the preparation of the Bird biplane for its record try. I was dismayed to find that Red did not share my enthusiasm for what I regarded as my golden opportunity to establish myself as a serious pilot. Unimpressed by the vague promises of G. P. Putnam, he not only refused to rejoice at the turn of events but promptly turned into a downright nag.

His pointing out the dangers involved in such a flight — takeoff loaded with high-test gas, forced night landing, fuel tank switching problems, meter failures with a heavy gas load aboard, blowing out to sea above dense cloud cover, and, last but not least, the question of my own personal endurance when pitted against the freezing cold of an open cockpit in a cloth-covered airplane – only stiffened my mulish resolve to prove him wrong.

Once I found out how strongly he opposed the idea, I felt I had to cover *all* the bases. Having already convinced Mother that this flight was essential to my future, I prayed all the harder that Father would stay on the road, knowing the tirade that would result if Red got to him first. (As it finally worked out, Father did come home two days before I took off, and Mother convinced him to let me go. I felt pretty sheepish that I hadn't thought of that solution myself.)

Despite his noisy misgivings, Red worked tirelessly with the ground crew, who were only too delighted to have another pair of expert hands at their disposal. His concern about the loaded takeoff was justified. Its problems were unique to this kind of flying and bore little relationship to any of my previous experience. It hadn't been long since the famous French war ace René Fonck, in a huge Sikorsky heavily laden for a transatlantic flight, had wallowed out of control into a fatal ground loop, resulting in an explosion and fire that killed two of his crew.

Red devised a series of sandbag load tests for me until everyone was sure I could ride the rudder bar no matter how heavily loaded the ship was. In the course of these trials I happily found that the Bird exhibited enough lift for two airplanes. Still, getting the tail up for take-off as I trundled down the field with a full load of gas would be the final test, so I tacked a wistful P.S. onto my daily prayers: "Please God, for January thirtieth, a stiff breeze out of the southwest. Just for a couple of minutes until I get off, OK?"

Red kept after me about my lack of experience with blind night flying, but even his ingenuity faltered in trying to simulate such a hazardous condition. Personally I equated being isolated atop heavy cloud cover at night to being trapped in a burning building – a disaster no matter how you looked at it – and no dry runs or fire drills could begin to approach the reality. On the other hand, I didn't believe in scaring myself to death by anticipating something that the odds were against.

Then came the news that Viola Gentry was planning to make a try for the same record. It caught us by surprise because the Curtiss crowd had deliberately kept it secret. Since there *was* no female endurance record, she could stay up for an hour and a half and still set one. Far from being worried, I was actually delighted. Competition like this would only enhance the publicity of my own flight, provided, of course, that Viola didn't stay up past the Bird's gas capacity. I admired her sincerely for making this attempt. She had had very little air time, and I doubted that the weather would be very cooperative.

Viola took off on December 20 around 6:00 A.M. and was back on the ground by 2 in the afternoon. The women's solo endurance record now stood at eight hours, six minutes. She had a world's record, and I had something tangible to shoot at. For once Red was all smiles, assuring me in a whispered aside, "That's just great! You can break that one standing on your head!"

In 1928 the need for radio communication between airplanes and the ground was only too apparent. But for those of us who flew alone, it remained just that — a need. Even if the equipment had been available, weight considerations would probably have ruled out carrying it. As it was, the whole load — including gas, oil, flight gear plus emergency signal pistol, food, thermos bottles, and all 112 pounds of me — had been calculated by the engineers down to the very last ounce.

The Bird was a nearly perfect craft for this kind of record. Once in the air, it was as stable as a flat-bottomed rowboat. I banked on this quality, knowing I could adjust the stabilizer as the fuel burned off so that if the weather were right, I could ride around the sky simply using the rudder. Even Russ agreed with me that in this respect the Bird held a slight edge over our beloved Wacos.

With an upper wingspan of thirty-four feet, the Bird was slightly larger than the Waco. Furthermore, the lower wing was shorter (twenty-five feet) and was staggered in back of the upper one, unlike the accepted biplane concept, which up to then had featured wings of similar lengths. And the Bird's wide-tracked oleo landing gear gave it a good solid under-pinning, far superior to the narrow axle type.

There were complaints from pilots about the Bird's "float" characteristics close to the ground, but I found that sideslipping and fishtailing cut the problem to a minimum. I was almost always able to set down within twenty feet of my target, and since airplane brakes were still a distant dream, this spoke highly for the air-plane's maneuverability.

Takeoff load tests were a snap. It left the post like a scalded cat and climbed like one chased up a tree by an enraged German shepherd. The Bird would last the distance all right, but the big question was: Would I?

On January 2, word came that Bobbie Trout in Southern California had upped Viola's record to twelve hours. This wasn't going to be the cinch I'd figured after all. The time lead I'd been banking on was narrowing. Unable to predict my own ability to battle the cold, I thought I could handle nine hours no matter what, but who knew if I could hold on for thirteen? (All records had to be broken by at least one hour.)

To add to my worries, the weather had been cold, still, and clear for days, and I wanted desperately to take advantage of it. But Bill Winkle wanted me to hold off until the last minute to prevent anyone else from waltzing into the air show with the record. He needed the prestige of this flight to lock in some promised financing, and I knew of all his problems in arranging that. The Bird was a superior airplane in all respects, and I resolved once more to hang in there and present it to the public in the best light I could.

Newspaper interest was heating up, and there was now a barrage of photo sessions and interviews to be handled. For the first time my relations with the press corps on the field suffered a strain. Telling them that I couldn't take off before the end of the month did nothing to relieve their impatience. A rip-roaring contest between female pilots was bound to be *news*, and they wanted the story now.

Viola showed no signs of making another try, so I could only assume the cold had done its work on her. But Southern California's mild climate just might induce Miss Trout to go up again. . . .

As the tension increased, everyone's fuse grew shorter. I confessed to Red that I wasn't sure I could take the freezing weather, but that I wouldn't settle for less than eighteen hours if it killed me.

"It might at that," he said sarcastically, "but might I remind you once more that nobody twisted your arm to do this in the first place!"

Remind me once more! That was about the thousandth time! By way of an answer I went around the hangar kicking tires.

Red phoned Father the night before the flight, saying that Roosevelt Field was in no state for a heavily loaded takeoff. Deep ruts cut into the ground during a two-day thaw had now frozen solid, and he feared a blown tire.

Father got on the phone and managed to move the flight over to Mitchel Field, a privilege rarely granted civilians. But Father and the commanding officer had become good friends the previous summer, after an overheating radiator had forced us to land there. Father's stories of his aerial mishaps had so amused the commander that he waived the regulations and let me fly the Waco out after a new radiator had been installed. (According to regulations, he was supposed to impound the plane, dismantle it, and have us truck it out!)

There was now a new CO on Mitchel Field, but he had been told about Tom Smith, and Father managed to charm him even faster than his predecessor. Permission was granted, and all signals were go.

January 30 dawned bright, cold, and windy. From New York's Battery Place weather station, James H. ("Doc") Kimball told Mother that it would remain cold and clear with diminishing winds. Good news indeed. At takeoff the gusts were so strong that the Bird sailed off the ground like a glider, despite the heavy gas load.

My custom-made flight suit was toasty warm, as were my fur-lined boots, gloves, and helmet. But the chamois face mask was troublesome from the start. Moisture from my breath kept building up and steaming my goggles. Having seen a Canadian bush pilot who suffered from facial frostbite by not wearing one, I was loath to remove it, although by the end of the sixth hour it itched torturously.

As night fell, the view of the island was spectacular. The weather was clear, with no moon, and the lights outlined the land like a diamond necklace on a jeweler's drape. But my delight at the scene below me was marred by the icy chill that had begun to seep through the layers of leather, fur, and wool encasing me. If the cold got trapped inside the suit, I would be finished.

Sundown was predicted for 5:50 P.M. That meant that I had probably been up for only four hours! At this rate I'd never make it. Doc Kimball's diminishing winds were tossing me around that wintry sky like a pebble in a wind tunnel, and there was no way the temperature could rise until they died down.

By midnight they were at gale velocity. I had loosened the seat belt a few notches to ease the numbness in my legs during the early hours of the flight, but the turbulence was so bad that even with my arms around the controls, I was in danger of being thrown out of the cockpit. This was a hazard I hadn't anticipated. I simply *had* to tighten that seat belt again in order to anchor my feet to the rudder bar. It took an eternity to get my fingers free of the fur-lined mittens, but I finally managed to tighten the belt and get myself firmly rooted in the seat.

This small triumph was instantly offset by my finding the stabilizer jammed or frozen. It meant that I couldn't trim the ship as the center of gravity shifted with the fuel's burn-off. In order to hold the nose up in flying position, I would have to keep both arms wrapped around the stick.

This lasted for about an hour — long enough for me to develop an agonizing rapport with those unfortunates who died on the rack — until one last desperate tug at the stabilizer handle jolted it loose. I wanted to cry with relief.

By three o'clock in the morning I knew I was finished, but I couldn't even contemplate making a landing. The truth was I had never landed at night. Everyone, including Red, assumed I had because I'd flown home many a summer evening with the lights winking on here and there in the dusk, but I was always safely back in the hangar by the time darkness actually fell.

One's depth perception becomes easily distorted in the dark. Therefore, sideslipping in at night with an airplane still carrying a load of high-test gas was too dangerous. Knowing this, why had I let them think I could handle it? Simply because I had not planned to end the flight at night!

Blaming my dilemma on Doc's high winds was no comfort. I knew exactly who was to blame for this predicament. *Me.* If I had told Red, he would have insisted on a rigorous period of training for night landings in good and bad weather, with and without lights, backwards, upside down, and sideways — you name it. Instead of being deeply grateful for this kind of solicitude, my successes in the past few months had swollen my ego to the point where it obscured the facts. I apparently thought I was a flier of such brilliance that I could mold the weather to fit my skills, but I was now faced with the unpleasant reality that I couldn't fly through the night.

I was wearing a parachute, but the thought of jumping over the side was more distasteful than the prospect of landing at night. How could I justify letting the airplane crash into the sleeping town below? Should I fly out over the ocean and jump there? The mere thought of a leap into the icy Atlantic left me colder than I has been in the first place.

I stared at the Véry pistol Red had insisted on clipping to the side of my seat. He had carefully explained how to use it, but in my bubble-headed enthusiasm to get started I hadn't paid any attention. Laboriously I flexed my fingers around the stick, trying to get enough circulation into them to hold the pistol. I told myself bitterly that the way my luck was running, I'd probably drop it over the side! But without the rocket signal from this deadly-looking blunderbuss, the ground crew would never turn on the landing lights. Why should they? Hadn't I told them I was invincible?

After a short period of flying, praying, wiggling my fingers, and puzzling over what to do next, I soon found myself back to staring at the Véry pistol. I had to give it a try. I pulled the throttle back, hauled the nose up, and banked steeply with my right arm hanging over the cockpit rim. I pulled the trigger, I thought, but nothing happened. No rocket, no light. Nothing.

Sheer desperation forced me to try again. I banked around steeply once more but this time shoved the nose down (where it should have been in the first place) and, by half standing up in the cockpit and using both hands on the pistol, managed finally to discharge it. For one panic-stricken moment I thought I'd set the plane on fire as the rocket lit up the sky under me.

The lights marking Mitchel's border blinked twice to let me know it had gotten the message. I later learned that the crew had had an ambulance standing by all night, but I never let on how wise that was!

In spite of my success with the pistol, the fact remained that I was still circling around at 1,000 feet on the coldest night of the year, terrified to land an airplane loaded with high-test gas. Suddenly I spotted the lights of an airplane beneath me. I watched the pilot circle into the wind and descend. There was moonlight now, and I could clearly see him line himself up behind two iced-over puddles gleaming palely in the silver glow. His movements were so deliberate that I could almost feel him yank the stick back as he set down smartly between the two natural markers.

Since he was landing at Mitchel, he had to be flying a military aircraft, which landed much faster than the Bird. I cut the throttle and started down. No fear of a sideslip now. My guide had given me such an accurate pattern that I didn't straighten out until the telephone wires were behind me. I yanked the stick back into my stomach as the tail skid and the wheels bit down hard into the frozen ground. The Bird seemed to be just as relieved to get down as I was, so I sat there for a moment letting the motor idle while I offered up fervent prayers of gratitude. I solemnly promised never again to display this blend of incompetence and arrogance. I would do plenty of other stupid things, but not this particular one.

I don't remember taxiing back to the hangar, but I must have. When I reached the concrete apron, I motioned to Red that I couldn't get out of the cockpit. The crew hauled me out bodily to be greeted by a frozen but wildly enthusiastic group of loyal friends, family, and newspapermen. I was glad to learn that in spite of everything, I had clocked a new record. Provided the barograph hadn't frozen, too, the new solo endurance record for women now stood at thirteen hours, sixteen minutes, and thirty-seven seconds.

I accepted all the acclaim with becoming, if hypocritical, modesty. Wild horses couldn't have dragged out of me what had really been going on up there. Attesting to my false bid for heroism, the thermos bottles that had been on the plane were found to be full of bouillon and coffee that had frozen solid. The papers made much of the "brave little girl flying through the night without nourishment," but I never told them why nourishment was the last thing on my mind.

Joe boosted me into the back seat of the family Buick and unzipped my suit and boots in order to massage some circulation into my stiff and aching limbs. Relief over the flight's success was evident in Father's light-hearted chatter from the front seat, although he warned that this was the last time he'd consent to this means of building a career. I was too groggy to do more than nod my head wearily.

Joe whispered to me, "That was some three-point landing, Sis. Nobody guessed you'd never done one at night before, not Mom or Pop, not even Red."

"All but you." I sighed gratefully at the heart-felt relief that he had known but hadn't told.

"Do you know who landed ahead of you?"

"No, who?" I asked, half asleep.

"Jimmy Doolittle."

"Jimmy Doolittle!" My eyelids shot open. Doolittle was our idol. His flying exploits had held us in thrall since we were children, and his scientific experiments in the development of blind flight instrumentation awed us now as adolescents.

"But why would Jimmy Doolittle..." I left the question hanging.

"Colonel Pratt called up at about eleven o'clock to ask how you were doing. When we told him that the temperature had dropped to eight degrees and the winds had increased to gale force, he couldn't believe you were still up. We figure he must have called Doolittle in Philadelphia before he left on an experimental night flight for New York. As I got the story from one of the mechanics, Doolittle was approaching Mitchel when he saw you fire the Very pistol, and he figured you might need some help getting down."

Talk about having a guardian angel! I sat silent, reveling in the knowledge that not only I had been rescued from a disastrous finish for the Bird and me, but I'd been saved from this ignominy by one of my heroes.

By the time I had thawed out and my hearing was fully restored Brunner-Winkle was demanding the air show appearance I had agreed to. It was delighted with the extent of the press coverage. According to the company I was now not only a "name" in American aviation but a household word as well.

"So's baloney," said my dear brother Joe.

At Grand Central Palace the air show was jammed and the Brunner-Winkle booth was crowded from morning till night. A seventeen-year-old girl record holder was as unusual as a cauliflower on a rosebush, so I drew a lot of lookers.

Brunner-Winkle got its financing, took orders for as many Birds as it could project production for, and I got the first of many telegrams that I would receive from Mayor Walker: "Glad you kept your word and stayed away from the East River. How about christening your plane *The City of New York?* Sincerest best wishes, James J. Walker."

6

G.P. Putnam Has Feet of Clay

After the punishing cold and turbulence of the endurance flight I found myself basking in the warm approbation of my fellow pilots. Everyone was jubilant, and there was much good-natured bantering about what records I would break next. I wasn't sure I liked that. I hadn't thought of myself as being in the record-setting business. All I wanted was honest work flying somebody's airplane for hire. And now with Brunner-Winkle's recommendation, I hoped it would come.

I sat back waiting for the phone to ring. I was particularly anxious to hear from G. P. Putnam, but Lady Mary Heath (who was in New York for the air show) insisted I should forget about that. "Mark my words," she said, puffing furiously on one of her scented Turkish cigarettes, "you'll never hear from him."

Her features disappeared behind a cloud of swirling smoke. "He thinks you deliberately outsmarted him that day in Chicago, and he'll punish you for it, never fear. And making a record like this, in full view of the press and public on the coldest day of the year in an open-cockpit airplane, is just rubbing salt in his wounds! Oh, yes, my dear, you'll pay for that, and dearly. Anyway," she added, "why do you feel you need him?"

Surprised at her vehemence, I answered slowly. "Oh, to deal with promoters, set fees for oil and gas endorsements, and — you know — *manage* everything. Mother hates doing it. Father is usually away, and I'm still a minor."

She looked thoughtful. "Well, maybe he *could* beat the bushes and come up with more paid public appearances than you can on your own, but as for the rest, you did as well this time in lining up endorsements as he could have, and now you don't have to pay out any fees."

I was about to explain that the endorsements had been possible only because Bob Loutt and Jap Gude, the local representatives for Kendall Oil and Tydol Gasoline, were personal friends, but I could see she was in one of her smugly righteous moods, the kind that brooked no contradictions.

At the time public interest in aviation was so great that I was paid handsomely for appearing at luncheons and dinners given by Rotary, Lions, and Exchange clubs. But I was bothered about being introduced as the World's Youngest Record Setter. I had learned the folly of being carried away by one's publicity after the bridge flight. The endurance flight, though, had not been a stunt. Its preparation and performance had been both rigorous and draining.

I wondered whether women like Trout and I were actually advancing the cause of feminism by these flights, as the journalists were now suggesting. I certainly hoped so. Beyond the vote, women in the twenties had few legal rights. This had always angered Mother. She did not feel personally put upon, but she railed at the prevailing assumption that women were second-class citizens. Since I had lived with this passionate revolt all my life, it pleased me that anyone thought I had struck a blow for change. I had just settled comfortably into my new identity when it was all swept away as if it had never happened. A few weeks after my flight Bobbie Trout, flying over Santa Monica in Southern California's balmy

151

climate, broke my endurance record by almost four hours! I couldn't believe that all my efforts had been for nothing. I forgot all about advancing the cause of feminism. Not for me the role of the sporting loser. I wanted that record *back*.

I welcomed Bobbie as a competitor and wired my congratulations. Now I had to find an airplane that coud beat her. But I didn't want to go through all that hard work and agony again, only to have the record rebroken in a matter of weeks. No, indeed. This time I would hang one up that would stay there for while.

For that I needed a much larger aircraft. The only planes capable of what I had in mind were Fokkers or Bellancas. I had met Tony Fokker at Curtiss Field several times. He was pleasant but aloof, and our conversations had been brief. I had met Guiseppe Mario Bellanca through Red and Martin down at Cross Bay. He had been impressed with my flying off the narrow sandy strip and had given me his card. "Get in touch with me," he had said, "when you are ready."

G. M. Bellanca was a legend in the aviation world long before Clarence Chamberlin flew Bellanca's *Columbia* to a nonstop long-distance record just a few days after Lindbergh's trip to Paris. Bellanca was revered for a variety of reasons that ranged from his personal modesty to the outstanding performance of his airplanes. Being told by him to "get in touch when you are ready" was tantamount to being invited aboard a space vehicle by Wernher von Braun.

Still, I couldn't be sure that it wasn't merely a gracious gesture. When he and Father first met down at Cross Bay, they had developed an instant rapport when they found that their mutual love of flight was compounded by ineptitude at the controls. They both also loved a good story and were polished raconteurs. I asked Father if he thought Bellanca was serious about his offer. Immersed in Dickens's *Bleak House*, he looked up quickly to say, "There is only one way to find out."

Up in my bedroom I stared at the phone like an enemy. Too much depended on this call. Bellanca's fortunes had vastly improved since we had first met two years ago, and he now had his own factory. I hadn't seen him since the previous summer. What if he didn't remember me? Smith isn't a very distinctive name.

I would be asking him to furnish me with one of his treasured planes and the use of his staff for the required engineering and mechanical services. The plane would need almost the same care and equipment as one slated for a transatlantic crossing, for this time I wanted to stay in the air for thirty-six hours.

The enormousness of what I wanted was compounded by the knowledge that Lindbergh had not even been able to buy one of these planes when he had the cash in his pocket. Still, Father was right. There was only one way to find out.

By now I was so agitated that I had trouble telling the operator to place the call. The connection to Delaware was hard to get. I was convinced it was a bad omen and had started to hang up when I heard the soft, familiar Italian voice.

"Miss Smith? Yes, of course I remember you. You are doing fine things in our business. I must say I have wondered when I would hear from you. Tell me, what took you so long?"

I couldn't ask him outright what I really wanted, but I didn't have to. He sensed it had to do with my last record's being broken so quickly, and he agreed that Father and I should fly down to New Castle to see what could be arranged.

I floated out into the hall and met Father. Reading my expression, he was as delighted as I was. This could be a quantum leap forward in my professional career. Bellanca kept a tight rein on his airplanes, allowing only a choice list of pilots in their cockpits. Everyone had to pass the scrutiny of his chief engineer and pilot, George Haldeman. The current piloting roster consisted of Clarence Chamberlin, Shirley Short, Martin Jensen, Roger Williams, Emil Burgin, and Bert Acosta — all experts. Even being considered to fly with this illustrious group was an honor. Compared to them, I was a rank amateur.

George Haldeman was a good friend of Russ's. I had met him at the time of his transatlantic flight with Ruth Elder in the fall of 1927. Headwinds depleted their gas supply, and they were forced down in the ocean some 300 miles off the coast of Portugal, but it was a courageous effort that gained the applause of the entire flying world.

Russ was delighted by my news, although he warned me that George would probably be harder on me than on a pilot unknown to him. I had to agree that this was only fair, but my heart sank just the same. I had heard that the Bellanca monoplanes were very tricky in their handling near the ground, although it was generally agreed that there was nothing around to equal their stability and grace once they were in the air.

Father and I set off for New Castle in the Waco, which he insisted on flying. It was a cold, blustery March day, and I silently decided that if we still had high winds in Delaware, I was going to wrestle the controls away from him, no matter what. I just couldn't risk his usual three-point landing with its blown tire, scraped wing, and broken propeller. Happily the winds died down, and Father made one of the best landings of his life, leaping out of the cockpit to make a low bow in front of G. M. Bellanca, awaiting us on the concrete hangar apron.

At lunch Father and Bellanca enjoyed each other's stories, while George and I compared notes on the logistics of the upcoming flight. That is, we did so after I found him treating the flight as a foregone conclusion. Apprehensive as I was about passing my qualifying tests, their confidence in me was too wonderful not to be enjoyed, even if it lasted no longer than lunch!

I was introduced to my new bird that very afternoon. Because I had never so much as sat in anything that size, her forty-six-foot wingspan and Wright J-5 225 h.p. engine made me apprehensive about my ability to measure up to Bellanca's expectations. Flying a little Waco 9 off Cross Bay's beach strip bore little resemblance to challenging this craft's throaty authority. But airplanes are like horses. They have to know and respect who's in charge. I did not dare be timid, although that's exactly what I was.

At George's signal I opened up the throttle, and the ground dropped away. If I thought the Bird climbed like an angel, this one had the inside track to the moon. Better still, she was more stable than any of the light biplanes I had flown. Landing her was tricky, but not for any of the reasons I had heard. I just needed longer arms. The throttle was in the middle of the instrument panel. The brakes were operated by the left hand; that meant that the stick was wedged between my knees and left largely unattended, while my feet busily operated the rudder pedals. Talk about rubbing your stomach and patting your head! Once I got the hang of it, though, it worked just fine, for the big monoplane had such a serene air of confidence that some of it rubbed off on me, and my beginner's jitters faded away.

The most wonderful thing about this airplane was that it did what I wanted it to before I got around to making the right moves. When I discovered this, my hands stopped fluttering between brakes, stick, and throttle, and the big ship invariably settled in for its own perfect three-point landing. A few more days of this, and our rapport was complete. We could now move on to more important matters.

Preparations began in earnest, and my days

were completely filled with load tests and gas consumption and engine trials. A pleasant respite (or so I thought at the time) was provided by a visit from Amelia, who flew in with Captain Bill Lancaster. I knew Lancaster well through Lady Heath. A celebrity in his own right, he was the first man to fly from England to Australia. He was also one of the reasons for Lady Heath's antipathy to G. P. Putnam. But I'm getting ahead of my story.

Amelia teased me about beating her to the "big ones," saying that she had planned to get there first. George Haldeman invited her to go up for a trial spin, and the three of us took off. She sat up front with him, and I stayed down in the last of the cabin's six seats. We had climbed to about 1,000 feet when George leveled off and motioned Amelia to take over the controls. Our big, calm bird suddenly lurged out of control and wobbled all over the sky. Amelia was embarrassed and motioned George to take over. He landed, and we disembarked in silence.

She pulled me aside and asked if we could go up again by ourselves. She was obviously as puzzled as we were. I turned back to the ship, explaining that only those pilots on the insurance list were permitted to take off or land the factory planes. She nodded understandingly, and we were quickly airborne. I flew about

15 miles north of the field. When I had climbed to 2,000 feet, I guided Amelia through some gentle, banking turns, carefully explaining the ship's normal flight position. We flew this one with the nose well below the horizon rather than on it, the usual position for small aircraft. I thought this position might have confused her before and played a part in her odd performance. It hadn't. Again we slipped and skidded all over the sky. I was baffled, for once the ship's flying position was established, the rest of it should have been exactly like flying a small biplane.

Actually the only difference I found in the big ships was in their handling near the ground. They usually landed faster, but I liked the increased sense of control that gave me — that is, in every large ship except the Lockheed Vega, which had all the glide potential of a boulder falling off a mountain. A three-point landing in a Vega often sounded like a small avalanche careening off the side of an Alp.

I pointed to the fuel gauge, and Amelia understood we would have to go back. I set down at the far end of the field to give her time to compose herself and me time to think of something to say. I have yet to live through a more awkward moment. Either Lady Lindy had never flown at all, or she had flown only briefly and quite some time ago. Today's performance didn't make any sense, except for Lady Heath's speculations that it was always Earhart's copilots (or "mechanics," as Putnam called them) who had done her flying. Now, finally, these remarks did make sense.

But there was no time to ponder this development. George Haldeman brought word that Louise Thaden had just taken off in California for an assault on Trout's record. Now I was really worried. Thaden was a competitor to be reckoned with. A fine pilot, Louise would come down only when the gas ran out. I didn't know what aircraft she'd selected for the job, but I comforted myself with the knowledge that she couldn't *get* a better one than I had. By the next afternoon I knew what I was up against. Louise had upped the record to twenty-two hours, three minutes, and twelve seconds.

Fortunately there had been few leaks to the newspapers about my absence from Roosevelt Field. Neither Ballanca nor I wanted Trout or Thaden going up again before we were ready. Walter Winchell made a few vague references in the *New York Mirror* to my "heart interest" in Delaware, but that was all. I assumed he was referring to my host in Greenville, Randolph Holladay, a wealthy sportsman pilot and friend of Bellanca's. Randy had installed an airstrip on his estate, which cut the driving time home after a busy day. But what most impressed me about his life-style was the fireplace in the bathroom of my suite. He did indeed seem to be the man who had everything, including Lana Holladay, a stunning blond bride who didn't believe the newspapers either.

When I finally left New Castle and flew the big bird back to Roosevelt Field, I was greeted by an ecstatic Herb McCory. "I knew you were up to something, kid, when Tom wouldn't talk. But I never dreamed it would be anything like this! Our little Ellie, flying the same ship as Clarence, Bert, and Martin ... the first girl to fly a big one! I don't believe it, but then again, I do believe it. Stand there by the prop, will you, while I get another shot? Oh, boy, wait'll the boss sees *this!*" The boss was Joseph ("Cap") Patterson, publisher of the *New York Daily News* and an avid flying buff himself. Both his daughters, Josephine and Alicia, were now taking lessons.

After another week's wait for good weather the big day rolled around. The barograph, swinging in its cat's cradle of wires, was installed in back of the gas tanks, which completely filled the forward section of the fuselage. I was carrying enough gas to stay up for thirty-six hours and break all existing records for sustained solo flight – long enough to take me across the Atlantic, should I want to go, although unfortunately I didn't yet know enough navigation to get out of sight of land.

Because of the heavy gas load, Bellanca decided it would be safest for me to use the transatlantic runway on the upper field. I had been standing on the edge of that runway two years before, praying that George Haldeman and Ruth Elder would get off safely for their flight to Europe. In my mind's eye I could still see George straining at the controls when the runway ended and the overloaded Stinson barely staggered into the air.

Happily today's takeoff bore little resemblance to that one. Given the same amount of gas and equipment, with less of a lifting wind (actually a cross wind), I was 100 feet in the air when I passed that point of no return. Compared to my first endurance flight, this one was a waltz. Seated in the comfortable cabin, I took none of the slipstream beating about the head that was so agonizing before. The balmy spring weather made it all the more enjoyable, for I had a constant stream of visiting planes flying up alongside me with messages printed on their sides. The hours slipped away like magic. I had been up six hours before

I even glanced at my watch.

Bellanca had taken no chances on my physical condition. He had sent me for a complete examination to a doctor who was also a nutritionist. I told him that in the past I had sometimes suffered bouts of airsickness during the monotony of the testing periods. To ward off any nausea that would impair my efficiency at the controls, he concocted some delicious raisin muffins for me to munch on and had thermos bottles of tea and ginger ale put aboard to wash them down.

Doc Kimball's weather predictions held good. It was a beautiful, clear, moonlit spring evening. I was even able to read by moonbeam. By four in the morning, though, my resistance sagged, and I became drowsy. I tried to snatch a few seconds of sleep at a time but found the temptation to keep my eyes closed too enticing.

To break the pattern set by monotonous circling, I decided on a little cross-country jaunt. Dawn found me almost 100 miles away out over the ocean off Montauk Point.

I knew that there would be much alarm in Garden City when I wasn't in sight, so I made a wide swing over the Atlantic and started back. I was met halfway by Mac in the *Daily News* plane with "WHERE IN H--- HAVE YOU BEEN?" painted in white letters on its side. I waved reassuringly out the window, and even from that distance I could see Mac's relief. I knew they were upset, but then they hadn't been droning around all night with me. My spurt of defiant independence came from knowing that as soon as I passed the twenty-two hour mark (Louise's latest record), I could never get out of gliding distance of the field. That point of no return would come at 10:00 A.M., so at 9:45 on the dot, I flew across the upper runway at 1,000 feet and from then on was never out of sight of it.

By noon I'd been up for twenty-four hours and had gotten my second wind. I was jubilant, absolutely sure I could make thirty-six hours. It would mean a night landing, but I'd be light on gas. Thanks to George Haldeman, I was no longer terrified of coming down in the dark. When I had confessed my lack of night landing experience to him, he promptly took over and taught me more about the glories and terrors of night flying than I needed or cared to know. Thanks to him, I was looking forward to tonight's set down with absolute confidence.

Trimming the stabilizer as the gas burned off had become an automatic gesture. But as the nose dropped once more and I reached down to turn the wheel, I found it wouldn't budge. One of two things had happened. Either the stabilizer itself had jammed, and I doubted that, or the cable that operated it and ran the entire length of the fuselage had somehow fouled. Whichever it was, I couldn't climb back over the gas tanks to find out because the ship was threatening to dive into an outside loop every time I took my hands off the controls.

I consoled myself with the thought that at least I had a two-hour lead on the record and that, with any luck, the pressure on the stick wouldn't increase too much, although from now on I would have to keep hands and feet glued to the controls and fly the ship every single minute.

An hour later I had to concede that Lady Luck was off at the race track. She certainly hadn't put in an appearance up here. Bellanca flew up alongside and saw me hunched over with both arms wrapped around the stick. He sent Mac back up with "WHAT'S WRONG? DROP NOTE IMMEDIATELY!" printed on his plane. Wearily I nodded, knowing that I should have explained the problem to them when it first appeared.

I wrote the note, wrapped it around one of the lead fishing sinkers I carried for that purpose, and aimed it down at the runway. I saw G.M. run out on the field, pick up the note, and trot back to the *Daily News* plane. In minutes he was back up alongside, waving frantically out the window, pointing straight down with one hand, and drawing a forefinger across his throat with the other.

I didn't want to land yet. I still had a sizable gas load aboard, and it was going to take an extra surge of adrenaline to give me the strength to yank the nose up into anything resembling a landing position. I knew the longer I could stay up burning off gas, the easier the landing would be. Still, when the boss speaks ... I dropped another note and kept circling. An hour later I knew I'd had it. I dove down past the gathering spectators to let them know I was coming in.

We touched down at 2:12 P.M., April 24, 1929, the Bellanca a gallant lady to the last as she bravely kept her nose as high as I could wrestle the stabilizer into letting her. The world's solo endurance flight record for women now stood at twenty-six hours, twenty-three minutes, and sixteen seconds. It still stands in the United States. Several claims were made against it in the mid-thirties in Europe, but they were disputed over the years, and the matter remained unsettled the last time I looked.

G.M. and Mrs. Bellanca were the first to reach the plane's side. They presented me with a bouquet of roses, the first of those I would receive from them each time I set a record.

The publicity again was beyond anything I'd imagined. One laudatory newspaper piece said I was seventeen going on thirty-five, a statement which might literally have been true if I'd added up the accumulated years of experience held by the pilots who had so painstakingly trained me. On the other hand, such efforts exact their price. If I'd failed to set a new record, I'd have had to leave town!

I enjoyed being a winner again, but this time I vowed not to make the mistakes I had made before. I had no way of knowing that the record would never again be challenged. So I continued to proceed as though Bobbie and Louise were waiting on the runway, throttles at the ready, all set to take off and snatch the record back.

Until the flight was over, I worried only about the flying. But my previous success now stood me in good stead when it came to negotiating endorsements and raising funds. I added Champion Spark Plugs to my list and got much more money than I had previously for gas and oil ads, although I refused to lend my name to any product I didn't use. Even clothing and fabric manufacturers wanted some of the fame to rub off on them, and I learned that the most persistent of these (a rayon tycoon) was the same one who had been wooing Earhart at the Chicago Air Show when Putnam asked my opinion about flying clothes.

The long awaited call from G. P. Putnam finally came. He suggested that we meet in his office on Forty-sixth Street just off Fifth Avenue. I was delighted. My dreams were coming true. Now, if he could handle all my bookings and...

During our meeting he was again impressive yet courtly as he asked me numerous questions about my flight. I answered politely but without elaboration. There would be no leaking of usable information today. I found that my initial shyness in his presence had largely disappeared. Maybe it was because of the news trucks riding around the city with my pictures plastered all over their sides, or maybe it was because I realized I no longer had to "prove" myself to him. With these two flights and contracts in my purse for $15,000 worth of endorsements, I knew that even if I weren't Lady Lindy, I wasn't doing too badly.

My inexperience, though, was no match for his polished suavity. If I thought I'd gained any leverage in dealing with him, I was badly mistaken. He had abandoned his plan for managing a group of pilots. "But you know, Elinor," he said, "I'm still very interested in your future."

My future, it seemed, was linked in his mind to Earhart's upcoming lecture tour which would follow the Powder Puff Derby (the first women's cross-country air race) in late August. "I fully realize the nagging financial insecurity that plagues most pilots and would therefore like to make you a generous offer," he said, bridging his fingers into a steeple and staring off into the space over my head.

"What would you think of a guaranteed seventy-five-dollar weekly income for a two-year period, — as a starting figure, that is? You would be Amelia's pilot and mechanic during the derby and the tour, and after that we could line up other duties for you, too. For the derby and tour, though, you would do all the difficult cross-country flying — A.E. is not physically sturdy, you know — but of course, she must appear to be doing it. When pictures are taken at various stopovers, you will see to it that you stand to her left, so her name will always come first in the captions. You will, of course, do no writing or public speaking for another two years at least. It's all spelled out in this contract." He flipped the papers toward me from across his desk, the hint of a smile playing about his thin lips. He was obviously enjoying himself, while I must have looked (and certainly felt) like a butterfly impaled on a pin.

Pushing the papers back to him, I said as coolly as I could, "I'm afraid I have other plans," and stood up to leave.

With a snarl he snatched up the contract and stood glaring down at me. "You may *think* you have other plans, but believe me, if you don't sign this, you will never fly professionally again and certainly never again in the New York area. You have my word on that!"

I forgot my promise to Lady Heath to be calm and reasonable. I don't remember everything I said, but part of my answer was that it would be a cold day in July before anyone could stop me from flying if that's what I chose to do. Then, with a slam of the door that shook the walls, I left.

By the time I collapsed onto the back seat of a cab my bravado had disappeared. I was appalled at what I had done. What a fool I'd made of myself! I dissolved in a flood of tears and told the cabby to drive just anywhere. I had to find a way to salvage my once-bright hopes.

It took only a short ride to convince me that I had no idea what to do next. I told the driver to take me to the New York Times Building in Times Square. Possibly Lauren P. ("Deke") Lyman, the *Times* aviation editor, would be there. I could trust him as a family friend to advise me on my chances for survival. As I told him my story, the tears started again, and this time I didn't try to stop them.

When I got to the part about being banned from flying, I said I was sure Putnam meant to ground me for the rest of my life. Deke smiled and handed me his handkerchief. "It's not funny," I wailed. "He'll do it. I know he will. Look what he's done to Lady Heath, Bill Lancaster, and Chubby Miller."

"Exactly what has he done to them?" asked Deke.

"Well, you remember last year when Lady Heath was the first person to fly solo from South Africa to England and Bill Lancaster and Chubby Miller were the first to go by air to Australia? Those were dangerous trips, particularly in those little Cirrus Avians they were flying. Each time Putnam welcomed them royally to America. Even before that Lady Heath sold her South African aircraft to G.P. and Amelia right after meeting them in London because she was impressed by Amelia's modesty and courage. Putnam gained the confidence of all three of them, and pretty soon the booked-up lecture tours they'd come to America to do disappeared over the horizon. Canceled — owing to 'lack of interest.' "

Even Deke was surprised by this revelation. He knew that far from being uninterested, the American public was clamoring to honor these Britishers.

"The worst of it is," I went on, "Lady Heath can't prove any of this since everyone involved is afraid of Putnam, but the head of the lecture bureau has just booked Amelia into all of her tour, and Lady Mary's livid."

"Why doesn't she sue?" asked Deke matter-of-factly.

"I told you. She can't prove it on paper, but she has good sources of her own here in America, and they all point the finger the same way. In any case, Lady Heath's husband would never agree to a lawsuit. He wants here back in South Africa, but it's totally unfair," I finished hotly.

"What about Lancaster and Miller?" asked Deke. "Do they feel the same way?"

"Of course they do, only they're broke and vulnerable. They're trying to get divorces so they can marry each other. Bill flew Amelia around last winter while they were still arm's length friendly with Putnam, but he was paid almost nothing. Now he and Chubby are stranded here without enough to get home to England. What Putnam was offering me was Bill's place!"

"Did G.P. tell you that?" asked Deke.

"Of course not. I just happened to know about Bill through Lady Mary."

"Do you remember if you mentioned all this to G.P. today?"

"No, I don't think so," I said slowly, "but only because I didn't think of it."

"Any word from Bellanca on the subject?" he asked casually.

"No, what has Bellanca got to do with it?" It was my turn to be surprised.

"You mean you don't know?" he asked incredulously.

"Know what?"

"Oh, my God! You remember when Amelia came down to visit you in New Castle before your flight? She had a check in her purse for twenty-five thousand dollars to buy a CH monoplane fully equipped for an endurance or long-distance flight."

"You mean she knew she was there to buy the airplane Bellanca was outfitting for me?"

"No, I didn't say that. She was told by Putnam that he was negotiating with Bellanca for an airplane and that she should have the check with her to close the sale if everything worked out. He then got on the phone with Bellanca and told him that he wanted your ship, since he knew it was the only one so equipped, and he had to have it right away for a flight he was planning for A.E.

"Bellanca called me to see if I could think up a way to get him off the hook and still not raise Putnam's hackles. I told him to call Putnam back and explain that he would have to consult with his board of directors on such a shift in policy. What I didn't say to G.M. was that twenty-five thousand dollars in hand was sure to appeal more strongly to a board of directors dominated by Henry B. Du Pont — another acquaintance of G. P. Putnam's — than the Bellanca Corporation's gambling on a seventeen-year-old girl to set a world's record.

"But Bellanca went me one better. He did as I suggested, and when the opposition raised its voice, he pointed out Amelia's flying record, which has been anyting but impressive so far, and convinced it that selling the airplane to Putnam would be the wrong move in the long run because if she cracked it up (which she undoubtedly would), G.P. would surely see to it that the airplane got the blame, not Amelia. On the other hand, the directors had been watching you fly around in it for weeks and up to then had been high on the prestige your flight would bring. And thank God, it all worked, and that's why Putnam got scared witless and made you that phony proposition today."

I stared at him, open-mouthed. "Why was it phony?" I asked. Deke went on to explain that Putnam was much too astute to think for one moment that if Amelia and I went on tour together, the press would be fooled about who was doing the flying. But if G.P. had managed to bluff me into signing that contract this morning, he could have kept me on the ground and bought the time that a long, drawn-out, messy lawsuit would have given him for the elbowroom he needed to employ his impressive scare tactics on any supporters of mine.

God had certainly steered me to Deke this day. His analysis of my near escape had my hair standing on end.

We went out for coffee and a conference with his close friend, Carl (C.B.) Allen, aviation editor for the *New York World.* The shock following the morning's revelations was wearing off. My feeling of being one against the world disappeared as I realized that I had at least two supporters. But the press had to be as wary of Putnam as we pilots did, for he courted favor among the highest of the publishing echelons, to which he had access from his inherited family wealth. He dined frequently with Adolph Ochs, publisher of the *New York Times,* and was often seen in the company of Herbert Bayard Swope, Frank Crowninshield,

Richard Berlin, and other prestigious news-paper editors and executives in New York.

The strategy developed out of that day's meeting was that Deke and C.B. would go as far out on their respective editorial limbs as possible to keep me in the air. This would be no small feat, for word had already been passed to them in a genteel but unmistakable way that Lady Lindy was America's top aviatrix, period.

This offended their professional sensibilities, but they had no choice, and they knew it. No matter what Putnam called it, it was a high-level form of good-old-boyism with all the members of his fraternity sticking by him. We now knew why no one was allowed to question who was actually at the controls when Amelia flew in for a visit.

This had long been a gripe of Lady Mary's. The first time she asked me if I'd ever actually seen Amelia at the controls when she landed at Roosevelt Field, I said, of course I had. It had been in the fall of 1928, several months after the *Friendship's* flight to Wales, when Amelia came to Long Island to be greeted by a group of Nassau County officials. She had stepped out of the rear of a single-engined Fokker Universal, been presented with a bouquet of roses, made a gracious speech of acceptance, and been promptly whisked off to a luncheon at the Garden City Hotel.

A man whom all of us knew to be a competent pilot had been clearly visible in the cockpit, but he had been identified by Putnam to the press as her "mechanic." The following day all the papers had carried accounts of Miss Earhart's "perfect landing." In reviewing the psychology that conditioned my thinking, I realized that I and everyone else there had *expected* her to be at the controls, and it had never once crossed my mind to wonder how she had disembarked perfectly coiffed, immaculate from head to toe, and wearing a dress in a plane that couldn't be flown in one! If I hadn't questioned her "perfect landing," why should a reporter, particularly one who had gotten the "word" from his editor?

It was a technique that had been developed over long years by Putnam in his career as a literary press agent. Far from being the serious publisher he presented himself as, I now saw that he really was: a publicist and impresario, orchestrating people's careers in order to peddle their life stories. My two friends at the *Times* and *World* described how for years they had seen him trumping up publicity for Putnam titles — books usually written by famous people, explorers, seafarers, aviators, and the like — the contents of which never quite matched the revelations promised in their advertising.

But even after all that I had now learned, I didn't doubt that Amelia was innocent of any complicity in Putnam's plans to buy the Bellanca I had flown. It would be completely in character for him to keep her in the dark about his schemes. Knowing full well she was too inexperienced to fly it, he would simply sideline the plane in a hangar until the day when she could. He would meanwhile line up backing for a future flight. With me out of the air, he could spread stories about my being too difficult to handle, about backers being forced to withdraw support, as he had already done about Lady Mary. It would be news. It would be published. And since I was helpless to refute it, it would be believed. Once again I was overwhelmed with the feeling of how precarious my career in flying was.

According to Deke, Putnam had already developed a foolproof response to any reporter daring to ask about Earhart's never flying alone into any of the metropolitan airports. Such inquiries were met with a frosty stare and the remark that Amelia Earhart was really too important to have to convince anyone. After all, wherever she went, her entourage followed.

C. B. Allen had learned, however, that she

was currently taking lessons, or at least a refresher course, at a secluded airstrip on an estate up in Rye, New York. He wasn't sure if it was Putnam's, but in any case it was close by. I was glad to hear that news. At least it showed she recognized the need for training, even if Putnam thought he could get away with his deception forever.

Warning me that I was still far from being out of the woods, my two friends made me promise to keep them posted, particularly if Putnam made good on any of his threats. Meanwhile, they would give me the same publicity as Amelia, provided I lengthened my current aerial track record.

This was important since she really didn't have one at the moment. They pointed out that playing David to Putnam's Goliath wasn't going to be easy and that we should keep quiet about everything we knew. For my part, there were to be no more outbursts and no more stunts like the bridge flight. From now on I was to become the most serious "girl" pilot on the eastern seaboard. That was fine with me since most of my competition was on the West Coast anyway. Louise Thaden was flying out of Oakland, Bobbie Trout was in Los Angeles, Marvel Crosson came down occasionally from Alaska, and Phoebe Omlie was flying out of some place in Arkansas. C. B. Allen warned me that Putnam was adept at freezing people out and that I should be prepared for the chill. Only Bellanca's personal friendship with Deke (something Putnam was unaware of) had saved me this time. We had to expect that from now on I would be turned down on flying jobs for being too young, or too blond and blue-eyed, or for parting my hair on the left side.

I was to keep both of them informed of my plans. I told them of having received a wire from Cliff Henderson, asking me to do a daily spot on his program for the upcoming Cleveland Air Races, for which he was Managing Director. I was placing that back to back with an offer to fly a Standard Oil Company Lockheed in the Powder Puff Derby from Los Angeles to Cleveland, and I was to do a radio broadcast for Standard Oil on NBC as early as next week, for which I would receive $1,500. If Standard Oil were as satisfied with the broadcast as it had been with my recent flight, it would back me in the derby.

On that hopeful note my friends and I parted. Two days later I got a call from Standard Oil about the radio broadcast. It was canceled, the caller said, owing to a mistake in scheduling. The company would be in touch with me soon about setting up another broadcast. I never heard from it. The arrangements for flying its Lockheed were never mentioned again either, no great surprise. But with that plane gone, I'd have to find another that could beat it, and that didn't promise to be easy. The Lockheed Vega was one of the fastest civilian aircraft around, but I'd need a lot of air time prior to the race to get the best performance out of it — assuming I could find one to fly!

Out of the blue I was offered a Vega by a rival oil company and made preparations to leave for California to get it. I honestly didn't like the Vega model. The cockpit was high off the ground, and your legs rested practically under the engine mount, making the heat build-up pretty uncomfortable at times. I'd flown it just once, and everything the Bellanca was, the Lockheed wasn't. Fast, unstable, tricky near the ground, with a monococque, no-longeron fuselage that would collapse like a box of corn-flakes in a crack-up, it would never win any plane-of-the-year awards. But it was the only plane fast enough to take the derby, provided it was flown properly. The oil company agreed that I could take home my prize money. I knew that Cliff was hoping to raise at least $10,000 for the race, so first prize would undoubtedly be $3,000 to $4,000.

Elated at my good fortune, I privately decided

that if I had to get out and push, this was one Lockheed that was going to roar in first over the finish line at Cleveland. This optimism evaporated as I sought to raise funds for the flight's expenses. As one door after another slammed in my face, even Deke and C.B. couldn't keep my spirits up. The only good news was a diamond watch that arrived from the Longine-Wittnauer Watch Company (I carried some of its chronometers on the endurance flight), along with a check for several thousand dollars. Miss Martha Wittnauer, an ardent feminist, had enclosed a card that said, "Please accept our expression of appreciation for your efforts in advancing the cause for women in aviation." As I graciously accepted her generous gifts, I didn't have the heart to tell her that the cause for women in aviation had just been stopped dead in its tracks.

I was still making the papers every day with what good copy was left in me, but as I ran out of things to talk about, I found myself leaning on the timeworn ploy of mystery. I said that all my plans had to be kept secret since all the other entrants in the race were keeping mum about theirs. While airily making these pronouncements, I wondered if everyone else was whistling in the dark, too.

When I had just about decided that George

Palmer Putnam was too tough for me, the phone rang. It was G. M. Bellanca, and after we passed the usual pleasantries, he asked, "How would you feel about taking one of our new Pacemaker models around the country on a demonstration tour?"

"You mean short of being crowned queen of England?" I shouted into the phone.

The Pacemaker was bigger than the Challenger I had flown for the endurance record. Even better, the Pacemaker had brakes on the rudder pedals, so the left-handed juggling act was a thing of the past. But best of all, she had a J-6 Wright engine of 300 h.p. that ran like a watch. If the Challenger was an aerial Rolls-Royce, this one was the new Silver Cloud.

I wouldn't be doing this for the Bellanca Company, G.M. explained. The plane had been sold to the Irvin Parachute Company, and I had been recommended as the pilot to fly it. I sensed that Deke had told him of my run-in with Putnam and that this was Bellanca's way of opening up the doors again.

He couldn't have picked a better way. When the news of my being hired appeared in the papers, Putnam was livid. He stormed into the Wall Street offices of the Irvin Company directors and demanded they replace me with Amelia. But my background and training were what had influenced Irvin most strongly, and the directors refused to back down.

I'd learned the technique of dropping jumpers from Red Devereaux and from Joe Crane, who was the leading parachutist on Roosevelt Field. My first demonstration for the Irvin officials up in Hartford, Connecticut, went so smoothly that two of them asked if they could ride back to New York with me. Always glad for company, I deposited them and the jumpers back on Roosevelt Field in record time, and there was no more talk of Putnam's getting the job away from me.

Before leaving on our national tour of civilian and military airports, which was scheduled to start in June, I flew the Pacemaker (which was nicknamed Gussie) back to New Castle for some last-minute adjustments. There was one fly in the ointment, and I quietly broached the subject to Mr. Bellanca, hoping he would understand. The rock-bottom truth was that I was distinctly allergic to parachutes. To me, this "sport" was akin to dropping a lifeboat into the open sea for kicks. The only way I would ever have left an airplane in midair was if the ship was afire and the wings had already peeled off. Yet I had every reason to feel that somewhere along this tour some enterprising reporter was going to suggest that I join in the "fun" and give him a great story by leaping over the side, too.

As usual, Bellanca came through in style by persuading the Irvin Company to have the tour's insurance cover me only as pilot of the airplane. When he hung up the phone and smilingly confessed that his feelings about chutes paralleled mine, my sigh of relief could have been heard back on Roosevelt Field.

7

The Show Must Go On

I had not seen Red Devereaux since his wedding in April, which I'd been unable to attend because of the intensive preparation schedule for the endurance flight. He had introduced me to his fiancée some months before, and I had found, much to my surprise, that I liked her very much. About five years older than I, she was lovely, had an impish sense of humor, yet possessed a serene maturity that belied her youth. Herma was perfect for Red in every way. It was with a distinct feeling of relief that I faced up to how one-sided my romantic fancies towards Red had always been and how lucky I was that he had been totally oblivious to my childish crush. I was also pleased to discover that his marriage had no effect on our friendship. His letters, filled with practical suggestions and constant encouragement, continued to come to New Castle and when the endurance flight was finally over, his joy at the outcome couldn't have been more evident.

I was troubled to learn of his appointment as manager of the new Syracuse Airport for the Curtiss Company. Undoubtedly this was because I was still inclined to view Curtiss as the enemy. (Curtiss people never stopped giving Father and all the other private pilots on the field a very hard time.) But then I hadn't thought anything could pry Red loose from his beloved Cross Bay enterprise. Obviously his new marital status demanded more stability and stature than the little beach-strip business could provide, and he was now proudly turning Syracuse into a topnotch airport.

In order to focus the eyes of the press on Syracuse, Red broke my light plane altitude record in a Curtiss Robin and jokingly suggested in one of his letters that it might be a good idea for him to take a shot at the solo endurance record in it. I recall asking somewhat testily how he planned to prop up the Robin's landing gear, assuming he could stash enough gas aboard to stay up that long. Without a lot of help that spindly undercarriage would be lucky to make it out the hangar door.

But then came the letter that gave me pause. He said he'd been racing a new high-winged Cessna monoplane at some local air shows and would be flying one down to New York on the upcoming weekend. How would I like to fly it with him, so he could show me what this baby could really do?

The answer to that was, not really. I'd already flown that model Cessna and was thoroughly disenchanted with it. I had recently had an experience at the Danbury Fair up in Connecticut which still bothered me. I was racing the Cessna at the time and was well out in front of the other contestants at the end of the last lap when I impulsively dove over the finish line. No sooner had I sailed past the judges' stand than the controls were seemingly wrestled out of my hands and the whole ship started shaking as though it were about to fly apart. Making a lunge for throttle and stick, I pulled back on the power, let her climb up gently on what momentum was left, and took stock of the situation. Once the engine slacked off, that terrifying shudder stopped and the aileron control became normal again. But there was no denying that I was in the middle of a pilot's nightmare — a serious structural failure while in flight. Either the ailerons were improperly hooked up, or the wings themselves could not sustain the strain of a dive at top speed. Either of these conditions would produce the wing flutter I had just experienced. The wings appeared strong enough to me now, but if the ailerons tore loose, I would have no lateral control at all. Gingerly applying just enough power to stay up without stalling, I circled around, watching the other contestants land. Only then did I come in on a long, straight, gentle, flat glide.

I said nothing to the judges who congratulated me on winning and made a ceremony out of presenting the prize money. They thought I landed last to heighten the dramatic effect. The drama that held my interest at that moment had nothing to do with winning the race, but of course, they would never know that. The man I wanted most to see – the Cessna engineer who'd asked me to fly the race for him – was pushing forward through the crowd. After motioning him to follow me into the hangar, I told him privately what had happened.

I had no wish to damage the Cessna Airplane Company, but I didn't want anyone else to go up in that aircraft again until it had been thoroughly checked out. The proper procedure, of course, would have been to make a report to the Department of Commerce and have it ground all these models immediately. But with the National Air Races coming up, that would impose a severe hardship on the company if this turned out to be the only airplane of theirs so affected. In my presence the engineer ordered the linen slit so we could see the top of the wing spars, which appeared to be in good condition. But there were other hidden factors to be considered, including metal fatigue, improper wing installation, and the design.

Arrangements were made that day to have the ship dismantled and shipped back to the factory at Wichita, Kansas, where it would be completely stripped down and subjected to the most rigorous testing. I was asked to put my findings in writing and address them to Clyde Cessna, the plane's designer and president of the company.

I did as requested, sent the letter off special delivery — and never heard another word. I might just as well have sealed it in a bottle and dropped it into the Atlantic Ocean for all the response I got. Not then or ever did Clyde Cessna acknowledge that "we" had a problem.

But I looked forward eagerly to seeing Red once more. His invitation was heartwarming and indicated that I had achieved a level of competence that placed us on a par. This was not true, of course. In spite of my record setting, I still had a long way to go to get into Red's ball park. Any time I didn't think so, Russ was right there to yank me back by the coattails, and if he wasn't around, there was Sonny or Pang or Bert Acosta. There was no way I could get a swelled head around that crowd.

Taking off from New Castle, I flew the Irvin Bellanca back to New York and met Red at Roosevelt Field. It was like old times – but better. There was no awkwardness in our conversation now. He was happily married, and I sincerely rejoiced in his good fortune. Over coffee at the field lunch wagon we discussed the upcoming Cleveland Air Races, an event that loomed large on the calendar of every pilot able to get there. He knew about my arrangement with Cliff Henderson to put on a daily acrobatics performance at the Cleveland Air Races and asked what I planned to do in my daily space on the program. Actually this had been nagging at me for some time. Henderson had every right to expect a sophisticated flying display, but this was obviously impossible in the heavily parachute-laden Bellanca.

Still, there were plenty of manufacturers of small, maneuverable airplanes who would want their aircraft unveiled in this manner before the national media and newsreel cameras. So essentially I was coasting until an offer from a company producing the best aircraft suited to the job turned up.

"Well, if that doesn't happen, how about the two of us putting on a mock pylon-shaving race in two of the latest-model Cessnas?" asked Red.

I took a sip of stale coffee (airport coffee was *always* stale) and said I'd like that a lot, only . . . and told him of my recent experience. He nodded soberly, saying that two other pilots in upstate New York had made similar reports. Up to that time he hadn't experienced it himself and was making no judgments until he did. My verdict put a different light on the matter, so he paid for the coffee, and we headed for the flight line.

He made a flawless takeoff in the shining red bird, climbing swiftly to 1,000 feet. Swinging left toward eastern Long Island, he dropped down over the Vanderbilt Motor Parkway and headed toward a tall electrical tower he selected as a pylon. Flowering dogwood branches clutched hungrily at our wheels, but he skillfully eluded their grasp. Skimming over an imaginary finish line, he grinned at me triumphantly. Then it happened — only worse than it had happened to me. Strong as he was, he couldn't hold the stick as it whipped violently back and forth between his knees. Yanking the throttle back, he brought her speed under control.

"Want me to try?" I asked hopefully. It was important to establish at just what point this frightening shudder would start each time. Wordlessly he unbuckled his belt, and I got into the seat from the other side. His try had been downwind. I decided to fly upwind to see if there was any difference. There was. It was worse.

After landing back at Roosevelt Field, we sat silently trying to organize our thoughts. Despite the gravity of this discovery, Red was reluctant to rock the boat. Curtiss was very anxious to promote the Cessna. The company was currently negotiating to become its eastern distributor. Furthermore, Red was still low man on the company totem pole, and professional jealousy was rampant in the ranks. Being the youngest as well as one of its best new pilots, he was currently scheduled to fly this model Cessna in the upcoming elapsed time city-to-city race that would start in Philadelphia and touch down at all the larger eastern cities before ending in Cleveland at the National Air Races. He had been so successful at the local air show races that Curtiss, in an unprecedented move (for it), had just agreed that he could keep half of his prize money. Presumably the prestige of flying for Curtiss was supposed to fill your stomach and pay your bills.

To insist now that the Cessna be put through an extensive scrutiny by the Curtiss Company (we had no way of knowing at this early date that Cessna itself would choose to ignore my initial warning) would do little to cement his relations with the company. Also, the fact remained that during his prior races, Red couldn't be sure that he had ever reached the speed we did that afternoon. Neither airspeed indicators nor tachometers were all that accurate. But he was most certainly going to have to dive across finish lines, hitting the same speeds as today, on his cross-country dashes to Cleveland.

To add to his woes, a particular hotshot pilot was breathing down Casey Jones's neck for Red's job. Fresh out of the U.S. Army Air Corps, this lad was an excellent pilot but left a lot to be desired as a human being. His pet ploy was to brand as "yellow" anyone refusing to fly any of Curtiss's obsolescent aircraft. He had been instrumental in the firing of two pilots I knew by secretly hiring one of our Waco mechanics to rerig and tighten up one of the ancient Curtiss relics and then proceeding to fly and stunt it within an inch of its life. Unfortunately the next pilot who applied for one of the vacated jobs was assigned to this same craft, which rather tiredly came apart in the air. Luckily the new boy wore a chute, so he was still around, but jobless. Somehow Curtiss managed to blame him for the ship's demise.

As if to prove Smith's Law (when things get really bad, you can always count on them to get worse), a glance out the plane's window revealed that Jones and the handsome flyboy were approaching. Slapping the cowling with a proprietary air, Jones asked, "Mind if we take over? I'd like Jimmy here to fly her around."

Knowing exactly how rough "Jimmy here" could be with any aircraft that gave him a chance to display his skills, Red told them what had just happened. Seeing their skepticism, I chimed in with my experience and pretty quickly wished I hadn't. It was obvious from Casey's disdainful expression that my opinion was unwelcome and unimportant. Before I did Red more harm than good, I got away with an invented excuse about seeing to the Bellanca.

Later, after the three of them concluded their conversation, Red came over and briefed me on the tricky fog conditions I might expect at Martha's Vineyard, the first official stop on the Irvin tour. He also said that Casey had promised to look into the matter of the Cessna's wing flutter. He did not say whether he planned to do anything about it or not. Red assumed he would. I wasn't so sure — but then there were a lot of things about Casey I wasn't sure of.

Gene Rock, Bert White, and I were about as disparate in nature and background as any flight crew could have been, but that was exactly what nurtured our mutual respect for one another's talents.

Gene was a handsome, fun-loving Italian with snapping black eyes and the physique of a bulldozer. Barrel-chested, he stood five feet eight inches and weighed in at 190 pounds of solid muscle — all of which he needed in performing his delayed-drop parachuting specialty. He was happily married, and his pictures of his wife and baby daughter were among his most prized possessions; he was never without a walletful of snapshots at the ready for instant inspection.

White, on the other hand, was tall, slim, and movie-star handsome. Catnip to the ladies, he was usually surrounded by a group of female fans hanging on his every word. With his blond good looks, smartly tailored uniform, and overseas cap set at just the right rakish angle, he epitomized the dashing young bachelor, a role he enjoyed playing to the hilt — provided it didn't interfere with business. Nothing was allowed to do that. Bert's various duties included participation in some of the jump demonstrations. In line with this, he set up lectures and classes at each of our stops. An expert chute packer, he taught military and civilian personnel everything there was to know about parachutes, learning in the process that the ignorance regarding this lifesaving device was often nothing short of appalling.

However, the "students" were not the only ones to learn something about parachutes they never knew before. Long familiar with dropping jumpers – and even an occasional delayed drop – I was totally unprepared for Gene's version of a delayed jump. To the uninitiated, a delayed drop meant jumping out of an airplane and not pulling the rip cord until the last possible moment it could be safely opened before hitting the ground. Since this was long before the advent of the easily maneuverable split parachute now used by sky divers, delayed jumping was a highly dangerous feat, one that could be done safely only by professional experts.

The first time Gene leaped out of the Bellanca, he hadn't told me it was a test drop, and I watched in terrified fascination as he plummeted toward the earth, a darkened speck against the landscape with no sign of the white parasol that would signal he was hanging safely below the shroud (was *that* ever an apt name) lines. He landed on his feet and running as he skillfully deflated the chute and gathered it up in his arms. He had asked me to climb up to 4,000 feet before I dropped him out. I figured he must have dropped 3,200 feet before pulling the rip cord. I could only guess at the shock of the chute's opening after a free fall of that depth. I knew by then that every landing presented a challenge to the jumper. Even a soft landing could turn into an instant disaster if a stray puff of wind managed to reinflate the chute before its wearer released his harness buckles. When that happened, the hapless victim was dragged relentlessly across the ground like a cattle rustler roped to the back of a sheriff's horse.

Despite Gene's expertise, there were still incidents that reinforced my allergy to chutes. I'll never forget the day in Ponca City, Oklahoma, when he landed outside the airport and lay still on the ground. I sideslipped down quickly and landed as near to him as I could. The impact of his fall had been so great that his chute buckles had sliced through the tough gabardine of his jump suit, cutting cruelly into his flesh and dislocating both shoulders. The shock of seeing him soaked in blood that he was trying to stanch with the white silk folds of his chute will never leave me.

But that was months into the future. Our tour for the Irvin Company officially started with our appearance at Martha's Vineyard, which happily was a rousing success. We were booked in as the star attraction for its local annual air show, which to Gene and me was like the clanging of bells to a couple of seasoned firehorses.

Assisted by some military jumpers recruited by Bert from goodness knows where, Bert and Gene were able to put on an exhibition of precision and formation jumping that whipped the press and spectators into an excited frenzy. Parachutes were nothing new, of course, but this type of exhibition had never been staged before, and I'm convinced that White and Rock were years ahead of their time with it. It certainly built the public's confidence that day, and demands for passenger rides became so insistent that a Ford Tri-Motor was flown in from Providence to handle the cash customers. It was part of my job to fly the VIP "freebies," but I soon found myself envious of the pilot of that Ford. Despite the fact that I was holding down a salaried job that pilots years my senior would have given their eyeteeth for, seeing him stuff that cash into a paper sack up in the cockpit brought back fond memories of my days at Cross Bay.

The main hazard on a day like this one was airport dust. This was the era preceding tail wheels and − in most cases − brakes, so practically every ship was equipped with a tail skid. The skid was a flat metal shoe that dug into the ground to slow your landing speed. It was perfect for grassy landings, but by the late twenties most flying fields had tracks worn into their turf where the prevailing winds had already laid out nature's landing pattern. A concentration of aircraft on a busy flying day could raise a dust layer as high as seventy-five feet that was as dense as fog and just as hazardous.

Under these conditions, tight-fitting goggles were a must to keep the seeping dust out of eyes, ears, noses, and clothing. After hours of flying through it, we all looked as if we'd spent the day in a mine shaft. That is, everyone but Bert White. He was his usual impeccably groomed self − nary a hair out of place.

"I swear to God he was born without sweat glands!" growled Gene under his breath.

Actually Bert's duties didn't call for our kind of physical effort, at least during our initial appearances. He was primarily concerned with planning our program and the public relations aspects of handling local VIPs, press, and visiting military purchasing agents. Unfortunately he had a blind spot when it came to the way Gene and I looked, or, as he put it, "the way your appearance reflects on the company."

Physically done in, we felt we'd given the company more than it asked for – and as for apologizing for the dirt that covered us from head to toe, our Italian and Irish tempers rose to the occasion, and the eggs really hit the fan. However, hot showers, a change of clothing, and an excellent lobster dinner prompted the dove of peace to spread its wings benignly.

The days passed quickly as we wended our way through New England's key cities and finally headed west to Albany, New York. We all were looking forward impatiently to the Cleveland Air Races. The papers of every city we visited were filled with advance publicity on what the public could expect to see, and if the show lived up to even half of these expectations, it promised to be the greatest spectacular the world had ever seen.

Everything from dirigibles to gliders was scheduled to be on hand, and the list of the pilots to fly them read like an aviation *Who's Who* compiled in dreamland. Long-distance flyers Charles Lindbergh, Clarence Chamberlin, Art Goebel, and Frank Hawks would be competing against speed kings Jimmy Doolittle,

Al Williams, Roscoe Turner, Doug Davis, Jimmy Haizlip, and Speed Holman. Stunt aces Freddie Lund, Frank Clarke, Holman, Doolittle, and Williams would vie for honors against crack U.S. Army and Navy stunt teams, a brand-new military concept in aerobatic performing that was being kept under wraps until the very last minute. The Powder Puff Derby had been on its way for days, and the women were racking up an impressive score marred only by Marvel Crosson's mysterious fatal crack-up, the cause of which was never discovered.

By the time we reached Buffalo the papers didn't seem to be carrying anything but news of the air races. Even the pictures taken of the three of us at Buffalo's airport were front-page news the night before we left, alongside shots of Dr. Hugo Eckener, commander of the *Graf Zeppelin* (a German dirigible which was now on its first around-the-world flight), and Admiral William A. Moffett, father of the navy air arm. The caption in one newspaper stated that we (Bert, Gene, and I) were the representatives of aviation's safety factor and, as such, would be importantly placed on Cleveland's daily program.

Overwhelmed at rubbing elbows vicariously in this fashion with such distinguished gentlemen, I was enough of a feminist to be delighted at my inclusion and thought back to when Mother finally became convinced of my determination and sincerity about aviation. "Be like the U.S. mail. Don't let rain, sleet, or storm deter you because you're a girl. If flying airplanes is what you really want to do, forget your sex and get on with it!"

I cut the front page off the paper and sent it home special delivery.

We came upon Cleveland from the lakefront in a cloudless sky. I circled low just once to let Cliff Henderson know we were there and then climbed back up to give Gene and Bert plenty of room for their jumps. Like the experts they were, they made perfect jumps and landed on their feet in front of the grandstand. Seeing them unbuckle their chutes, I dove down over their heads and pulled up into a climbing turn directly into the wind. Newsreel cameras were on the flight line, and our Irvin insignia – *Happy Landings,* with billowing parachutes painted on it – was displayed along the fuselage to full advantage (making the Irvin officials present very happy).

Continuing to climb, I hung Gussie on the prop, going into a half chandelle before cutting the throttle and kicking the rudder to hold her steady in a steep 800-foot sideslip to the ground. Straightening out, I hauled the stick back into my stomach, and the ship did the rest, settling in as gently as a hen on a nest of eggs. I didn't even have to apply the brakes. She rolled to a stop all by herself. All I did was cut the switch. Gene, Bert, and Cliff yanked the door open and lifted me out. We three, with Cliff beaming proudly, took our bows in front of the packed grandstand.

The roars of thousands of excited fans reverberated like a pounding surf. They obviously wanted an encore, but with so many events tightly scheduled, we begged off via Cliff's announcer up in the stands. We really couldn't have done it anyway, despite our enjoyment of this overwhelming display of appreciation, because none of us had had any breakfast. Buffalo's hotel coffee shop wasn't open when we checked out, and there had been no other opportunity since. When Cliff learned this, he steered us into his office beneath the stands and sent out for some food. He left us there, saying we wouldn't be disturbed until we'd eaten, but he had to get back on the field.

We had barely finished out sandwiches when he was back, his face gray. Wordlessly he handed me a telegram: "Red Devereaux and wife killed in dive Boston Harbor.

Ailerons tore off Cessna. Tell Elinor. Tom Smith."

I stared at it unbelievably. Killed? How could that be? The maps he'd painstakingly marked out for me for the eastern portion of the tour were out in the cockpit ... he'd told me to keep them as long as I needed them ... but — how could this have happened ... the wing flutter ... the ailerons ... God *damn* them ... they said they'd check them out ...

Gene tried to comfort me, but I tore myself away. I was in a savage rage that I didn't understand. What I knew was that this didn't have to happen. This was no act of God's — and it certainly wasn't Red's fault. It *could* and *should* have been averted. Suddenly I doubled up with a pain that tore at my vitals. I had never known emotion like this. It was grief, anger, outrage, self-condemnation, and regret all lashing me at once. I couldn't cry, but I ran blindly like a wounded animal until I found myself in the back of the darkened hangar, staring up at an aging Waco 10 someone had left there. The sight of it brought back so many memories I was lacerated by their pain. Mercifully this new agony released the tears, and I cried my heart out.

When the emotional storm passed, I realized someone had found me. It was Ted Van Deusen, an old friend and ally. Formerly a newspaperman, he was now doing public relations for Pan American. Van pointed out gently that an announcement would have to be made from the stands if I canceled today's performance and that Cliff had just received word of two other cancellations from today's program owing to engine troubles. If I felt up to it, they would like me not only to drop the Irvin jumpers but somehow to extend my time in the air to cover for the other two dropouts. What did I think?

I looked at him dully. "Right this minute I *can't* think, but if you can come up with something to fill the time, I'll give it a go."

Ted hugged me, saying, "Good girl! I told Cliff you'd come through."

Actually I couldn't think of anything I wanted to do less, but I knew my private grief could not intrude on my professional obligation. The Irvin Company had paid for a spectacular exhibition, and that's exactly what Bert, Gene, and I planned to give it. But as to extending our air time, I was too drained...

Van had been nervously pacing up and down. Suddenly he spun around and exclaimed, "I've got it! You've been to the Barnum and Bailey Circus, haven't you?"

I nodded wearily. At times Van got carried away, and I hoped this wasn't one of them.

"Well, remember that stunt where a big limousine is driven into the center ring and when the doors open dozens of midgets tumble out?"

"Uh-huh, but what have midgets got to do with...?"

He was excited now and not listening to me.

"How many jumpers have you dropped at one time on this tour?"

"Oh, I don't know — two or three."

"How many people can you jam into the Bellanca and still get it off the ground?"

"Well," I said slowly, "if we take the seats out, maybe seven — that's eight, counting me."

"If we roll it into the hangar, could you take off right out of the door?"

"I guess so, but the prop blast will sure make a mess out of anything behind me."

"That's not important right now. The element of surprise *is.*"

Unnoticed by us, Cliff was standing by, and Ted turned to him now, saying, "Get White over here on the double, and tell Pan Am I want a couple of their best mechanics right now! I also want some bolts of black cloth, tacks, and a hammer. We're going to have to block off those windows..."

Cliff turned without a word, and Ted led me back to the office. For the next half hour there was much bustling about as jumpers appeared out of nowhere and mechanics wrestled the seats from their moorings, while I wrung my hands and prayed we could get them back in again when all this was over.

They finished ten minutes before we were scheduled to appear, just time enough for the jumpers to sandwich themselves inside the fuselage after I squeezed past them to get into the cockpit.

I started the engine as the huge hangar doors were rolled back. Standing on the brakes to get the tail up into takeoff position, I slammed the throttle open, and we flew straight out of the hangar door. I was braced for trouble, but I needn't have been. Gussie sailed upward like a soaring sea gull — to the great relief of the jumpers. Veterans all, they knew that a cross-wind takeoff under the best conditions could be tricky, but with an overload like this, on a hot, muggy day with lift at zero, tricky could turn into disastrous at the flick of an eyelash.

White then choreographed what turned out to be the first mass parachute drop in history with the precision of a military drill, timing each jumper with a stopwatch. By now I'd become adept at getting the ship's tail out of the way for each jumper by pulling back on the throttle when he jumped so he wouldn't be rolled around in the propeller blast. What I failed to figure on for this mass jump was the loss in altitude each time I throttled down. With two or even three men I could hold the loss in height to about 500 feet. With seven jumpers it was closer to 1,000 feet despite my most determined efforts. White, the last man to jump, went out the door at an altitude of less than 800 feet! (The minimum *should* have been 1,000 feet.) Bert may have looked like a movie star, but under pressure he never failed the parachute business.

While I circled the jumpers on their way down, the tears started up again. Wiping them away, I suddenly heard, "Come on, none of that now. You *said* you wanted to be a pro — well, this is all part of it. Now don't disgrace us all by making a bum landing in front of that grandstand..."

Startled, I looked around. The cabin was empty. But I knew that voice ... and I knew I'd never see its owner again.

8

Off to the Air Races

The arrival of Deke Lyman and C. B. Allen to cover the races for their respective papers did much to ease my grief, for in their presence I could talk out the emotional trauma I was working my way through. In the late twenties people still liked to believe that aviators were a breed apart – a gallant group who ignored or turned their backs to the tragic events inherent in their calling, yet flew bravely on, daily facing their own doom. My practical newsmen friends paid no attention to this nonsense, realizing that this legend had been around since the time of the Wright Brothers and had as much basis in fact as the other timeworn myth that all lady pilots were lesbians because they wore shirts and pants.

They knew we kept right on flying for the same reason they kept on news gathering — because we loved it and because a lot of us couldn't make a living doing anyting else. As a flying officer in the National Guard, C. B. Allen encompassed both worlds. An excellent flier himself, he was quick to recognize piloting ability (or the lack of it) in others. But he, like Deke Lyman, never let his personal feelings interfere with his professional judgments. You might not always like what C.B. wrote about you, but you knew you could rely on his bedrock veracity, something I was coming to lean on more and more.

The one thing I couldn't bring myself to tell them about was Red's voice that day in the Bellanca's cabin. Tolerant as they were, I feared they'd think I was over-dramatizing or just imagining something that hadn't happened at all. I *knew* that voice was Red's. He had a funny way of chopping out his words that was unmistakable. But in retrospect I, too, was puzzled about how I'd heard him. He'd talked to me in a normal conversational tone, not with raised voice to be heard over a motor. Had I heard his voice *inside* my head? If not, why hadn't he been shouting?

It would be a long time before I got the

answer to that. Meanwhile, I decided to keep it to myself. Life in those days was defined in black and white. If you knew what was good for you, you stayed far away from the grays. I had no wish to join the ranks of those already escorted to a rubber-lined room by some men in white coats.

Van had been dead right about his "element of surprise." The mass parachute jump turned out to be a sensational blockbuster that we were asked to repeat every day for the balance of the show. In addition, I found myself performing another pleasant task — flying as glider tow pilot for Frank Hawks.

When Frank first asked me to use the Bellanca to tow him, I was intrigued with the idea. Then under contract to the Texaco Oil Company, he was making an aerial tour of the United States by glider. Frank was currently the holder of the nonstop transcontinental speed record, so the reasoning of some advertising genius in having him pilot a glider towed by someone else completely escaped me — as it did Frank. But he said the peace and quiet of the glider after roaring back and forth across the United States at full throttle were so soothing to his nerves that he had to pinch himself to keep awake.

Maybe so, but not, I'm sure, while riding in the prop wash of Duke Jernigon's Travel Air. One solo in that glider convinced me that Frank was pulling my leg.

When Frank cut himself loose from the towline, he always circled the field, performing a few graceful aerobatics before gliding in to a perfect precision landing in the circle marked out in front of the grandstand. It looked so easy when he did it that I was sure it would be a snap, but when I released the cable connecting the glider to the tow plane, I found that all my gliding around in powered aircraft hadn't prepared me for this. It was quiet, all right, but I quickly discovered that I'd best forget about any fancy footwork and make sure I didn't stall this feather-in-the-breeze. I shoved the nose down, leaned on the stick to keep it there, and stamped both feet as hard as I could on the rudder bar. Forget about making a precision landing! I had to fight to get down on the field at all, as the rising heat kept lofting me higher and higher. I finally settled in so far out on the edge of the airport that they had to send a truck after me to bring the glider back!

The very next day Duke came down with appendicitis, and Frank thought this a good time to put the Travel Air in for a complete engine overhaul. Using the Bellanca to tow the glider didn't present any problem, except for the cable reel-in. That would have to be done by hand, down in the rear of the cabin. Bert White rounded up one of his ever-present volunteers, and we put in a make-shift wheel for him to use. It seemed safe enough to Frank and me. To us the most important thing was getting him into the air on schedule. The daily aerial program had to be rigidly adhered to by the regulars. If it wasn't, Cliff Henderson wouldn't have the time flexibility to present the number of visiting celebrities like Lindbergh, Al Williams, James Doolittle, and others who hadn't, for various reasons, been able to assure him of their presence before they turned up. So far everything was going smoothly. The show was an enormous success, drawing crowds in excess of 100,000 spectators, who overflowed from the field and grandstands out over the surrounding countryside.

In making the switch from the Travel Air to the Bellanca, neither of us took into account that the Bellanca's J-6 engine was double the horsepower of the Travel Air, so when I opened up the throttle, I blew Frank a couple of hundred feet in the air, where he hung helplessly with no control over the glider. Slowly reducing power, I brought him back down unharmed, but we knew we were now faced with a serious problem. The only way I could get him into the air safely was to take off at just under half speed. This wasn't much above my stalling point, and I wasn't sure it could be safely done. But Frank was willing to try, and the Bellanca had yet to fail me in a crunch. Nor did it that day, coming through with flying colors.

We adopted this procedure from then on, and all went well until the last day of the show. Frank was already cut loose from the lowline and I was circling him on his way down when a sudden yell from my cable reeler made me look around just in time to see the heavy wire loop up off the cylinder and jam itself firmly down into the tail section. This left about 100 feet of steel cable arcing out behind us with a huge hook dangling at the end of it. Catching ourselves on any obstruction while in flight would be disastrous, but Cleveland, like most major airports of that day, was outlined by a lethal crisscrossing of uninsulated high-tension electrical conduits. If I hit those, half of Ohio would go up in smoke. Needless to say, Smith's Law went into immediate action. A glance at the fuel gauge showed the needle resting on E − as in *empty*.

Now on the ground, Frank saw what had happened. I watched aircraft being yanked tail first into hangars and people scurrying around as Frank bellowed for them to clear the field. I climbed up another thousand feet to gain the altitude I'd need when the engine went dry — as it abruptly did a few minutes later. After throwing Gussie into a long, steep, sideslip, we landed smack in the airport's center, the steel cable writhing behind us like a giant snake.

C.B. was one of the first to reach us with the gas wagon. Leaning into the cockpit, he sighed heavily. "I can't believe you did this on purpose, and I *know* Hawks wouldn't, but will you please tell me why, whenever there's a cliff-hangar finish, you always turn up in the middle of it?"

Pretending a bored yawn, I shot back, "Just lucky, I guess," and we both howled.

Russ Holderman came roaring in, having been told by a disgruntled bystander, "That Smith kid's grandstanding again — made a dead-stick landing in the middle of a race with Hawks's tow cable hanging from the Bellanca like a hundred-foot tail...." I found out later he silenced this critic with a single forceful word that begins with *B* and ends in *T*. But he wasn't letting me off the hook without a detailed explanation.

True, I explained, there was a race going on, but the pylons were nowhere near the center of the field, so that part was all right with Russ.

But getting caught without an adequate gas reserve was something else. It took C.B. to point out to him the impossibility of getting gas trucks to service us as we kept to the daily schedule. Actually, if the cable hadn't jumped the tracks, I still had a few teaspoons of fuel to make it back. I'd done it several times before. I left Russ with C.B. and went off to my parachute-dropping chores. When I got back, he was all smiles and promised to arrange a meeting with Jimmy Doolittle as a peace offering.

Earlier in the week the arrival of the Powder Puff Derby had put a big dent in my plans to become America's number one aviatrix. When Louise Thaden roared across the finish line and slid that big speed-wing Travel Air into a smooth-as-cream landing, my hopes plummeted. This wasn't going to be the piece of cake I thought it was. Up to now I had had the heavy plane field pretty much to myself. The only ones to match me in flying expertise were Lady Heath and Phoebe Omlie, both experts in the 90 h.p. class who had no plans to move on to anything else.

In pursuit of my goal I had flown everything I could get my hands on, military and civilian alike. I'd soloed everything from Aeroncas to

Ford Tri-Motors, from Vought Corsairs to Curtiss Falcons, but I was totally unprepared for the ease with which Gladys O'Donnell came in behind Louise, sideslipping her big clipped-wing Waco to a feathery three-point landing. These ladies might not have graduated to cabin ships like the Bellanca or Pacemaker, but they surely knew their way around heavy biplanes, and if I didn't hotfoot it to some unusual skill building nobody'd thought of yet, I would find myself out of the running.

A shout went up as Amelia's Lockheed hove into view, but when the elapsed times were announced, she finished a full two hours behind Louise, a disappointing third. Because she was flying the fastest ship in the race, her lack of expertise in both navigation and flying was pitilessly exposed. Her landing in Cleveland was amateurish as she bounced the big monoplane completely across the vast airport. There were snide remarks from onlookers when she frantically braked the ship out of a ground loop before rolling to a stop. But at that moment I was filled with admiration for her. Had her detractors known what they were looking at, they would have been cheering.

It was barely five months since the New

Castle incident. In that short time Amelia had obtained her private license (because of Putnam's careful statements, few were aware that she hadn't had this all along) and made the transition from solo student to racing competitor. There was absolutely no way she could have built up enough air time in that brief period to be at ease behind the controls of the fastest heavy monoplane in the air. I could envision her terror each time she made a landing approach, knowing that she must keep her speed up or drop into a fatal stall. The landing speed of the Lockheed was at least a third faster than any other plane in the race, and some of the airports along the way were little more than cow pastures. Her difficulties were compounded by the extreme heat. When she failed to realize that this would thin the air, making it less resistant, thereby decreasing its lift and *in*creasing her landing speed, she nosed over in Yuma, Arizona, and a new propeller had to be flown in to enable her to stay in the race. Despite this delay, she flew on to overtake the others but used up more precious time when she lost her way on several laps. From Putnam's standpoint, her entry into this competition was a disaster. But from hers, it was a challenge that she met head-on.

Lyman and Allen shared my admiration for

her performance, knowing we were watching a drama being played out that couldn't be shared with anyone else. Putnam's power precluded that, but one look at her drawn countenance when she flipped up the cockpit hood told us we were righter than we wanted to be. This was gut courage that transcended the sanity of reasoning.

Much later I heard through the New York grapevine that Putnam was acidly critical of her performance in the race — but only to her. To everyone else, he blamed her poor showing on the airplane, saying Lockheed hadn't put it in proper shape before she left Santa Monica.

She called me several days later, having just learned that women were being excluded from the big-money prize races. She was toying with the idea of forming an organization for women pilots as a first step in combating this form of discrimination. Over lunch she pointed out the paltry amount of money raised for the Powder Puff Derby (about $9,500) and how it bore little relation to the amount which poured into the pockets of the local promoters of the shows where the women appeared.

She asked me bluntly what I received for stunting and racing exhibitions. When I told her, she seemed surprised.

"Do you ever have trouble collecting it?"

"Not so far," I answered, "but that's probably because I don't have any competition."

"What rule of thumb do you use in setting a fee?"

"The same as G.P. uses in setting yours — the size of crowd I feel I can guarantee. The only difference is that I have to perform for it."

"Touché!" She smiled wryly. "But I have to guarantee attendance to my chicken-and-peas dinners."

"I'm not so sure of that," I said. "G.P. is a powerful persuader, *and* you don't have to fly for your money."

"That's been a big mistake," she said ruefully. "I should have more than enough air time by now to be doing it, but he's got the idea that it's demeaning somehow ... and this Lady Lindy thing ..." She shook her head in exasperation.

Much as I liked the idea of an organization for women pilots, I couldn't go along with hanging it on our entry into big-money races ($5,000 to $10,000 in prize money), like the Thompson Trophy or the Pulitzer, and the "free-for-all" races (where anyone who wanted to could compete for $1,000 to $3,000 in prize money) purely on the basis of safety.

"Amelia," I said (I could never bring myself to address her as A.E., feeling this an affectation of Putnam's), "let's look at the facts. Of the twenty entrants in the Powder Puff, how many do you honestly think could hold their own against Jimmy Doolittle, Frank Hawks, Roscoe Turner, and Al Williams?" Without waiting for her answer, I went on, "Assuming they could get sponsorship and access to high-speed aircraft, the only ones qualified to fly them would be Louise, Gladys, Lady Heath, Phoebe, Mae Haizlip, Thea Rasche, and yours truly."

She bristled visibly in defense of the other entrants, saying hotly, "Just because they've never done it doesn't mean . . ."

"That they can't? Of course not, but it's taking a wild chance on everyone else's safety — to say nothing of their own."

She frowned. "Granted all you say is true — how do we right this situation?"

"I didn't say I knew how to do that. You're the one with the administrative skill . . . but if we concentrated more on establishing women's races, until the day comes when your Powder Puffers are willing to compete on *qualified* equal footing, I think we'd stand a better chance. Anyway, why not turn G.P. loose to raise prize money for the first one?"

"Because he wouldn't do it!" she said with unexpected asperity. At my surprised reaction, she went on, "He's not very big on labors of love – or hadn't you noticed that?"

I chewed on that one all the way out to the field. Was she getting tired of being the money tree? Had it just dawned on her how completely she was being exploited right down to the short haircut and the publicized Lindbergh resemblance?

Of all the entrants in the derby, Bobbie Trout had the most unfortunate series of trials. A forced landing in Baja, California, wiped out the landing gear of her small Golden Eagle Kinner monoplane, and the delay in getting another one to her forfeited any chance she might have had to finish in the money. Nonetheless, she doggedly flew on to make a triumphant landing in Cleveland. By that time her sponsor, feeling sure she would give up and turn back, had already given up her hotel room to someone else.

Knowing that I had comfortable quarters arranged for by the Irvin Company, Phoebe Omlie brought us together, saying that two such determined competitors should get to know each other. Luckily we took to each other immediately, so that problem was solved.

But this started C.B. worrying about the lesbian legend all over again, for Bobbie epitomized the high-fashion concept of the day. Tall and lean, she wore her hair cropped in the very latest boyish bob, with pointed sideburns. Because she had no hips or bust, her gabardine trousers and tailored shirt fitted perfectly. I tried to explain that she wore boots because of the desert snakes and in the middle of it broke down in giggles because at the time I was wearing almost identical ones whose only value to me was to look sharp in the newsreels. As for Bobbie's sexual preference – all I know is that around me she was always a perfect lady, and, as I got to know her, a most courageous and gallant one as well.

I was crushed to discover that Lieutenant James Doolittle wouldn't be appearing at the races after all. I had hung my hopes on meeting him – or even getting to see him at close range – on Cliff's press releases and Russ's promise. But the word came down that his work with the Guggenheim Foundation was too pressing to permit him to leave Long Island, and that, evidently, was that.

The daily stunting exhibitions went beyond Cliff's wildest expectations. It got to the point where you didn't believe what you were looking at. The Navy High Hat Squadron of nine planes performed the most intricate aerobatics while *roped together* in units of three — flying triangle formations. These fliers were so expert that during the whole meet only one rope parted on a takeoff. And Speed Holman had people on their feet by looping and spinning a Ford Tri-Motor, its metal skin glistening in the sun. This was a very heavy plane, and such stunting had never been done in it before. When one envious flier made a remark that any hot barnstormer could get away with a thing like that *once,* Speed went back up and did it all over again!

Charles Lindbergh, now officially a colonel in the U.S. Army Air Corps, flew in with his bride, Anne Morrow. He had been invited to view the proceedings from the official military box but announced he'd rather be a part of them, and the next thing anybody knew he'd climbed into a pursuit ship that was totally unfamiliar to him, and with two army buddies proceeded to stunt the High Hats out of the sky. It was the first time I'd seen the squirrel cage loop done, and I was as awestruck as the rest of the paying customers. Seeing three fighter planes chasing one another around in an enormous loop about 1,000 feet deep without faltering, even though they were all flying directly in each other's prop wash, was a lesson in control from which we all could learn.

But then came the announcement over the loudspeaker that made my day complete. As a result of a personal request by Major General James Fechet, head of the U.S. Army Air Corps, Lieutenant James Doolittle would make an appearance at the races after all.

He flew in from Long Island the following day in a Corsair. A supercharged Curtiss Hawk had been readied for his aerobatic display, and he immediately switched planes for a test run in the Hawk. We all were hoping to see him do his famous outside loop, although the army had banned it as too dangerous after he first accomplished it two years before.

Doolittle climbed quickly out of sight in the Hawk, and we presumed he had gone out over the lakefront to practice. Within half an hour he was back on foot and jauntily carrying his parachute under his arm. During a practice dive the wings of the Hawk had peeled off, and he had parachuted out! (This automatically made him a lifetime member of the "caterpillar club," an unofficial society of people whose lives had been saved by the silk of a parachute.) Nothing daunted, Lieutenant Doolittle borrowed another ship and went back up to put on a breath-taking exhibition of inverted flight that opened the eyes of the army, navy, marines, and civilians alike.

I did get to meet him, as did a couple of hundred others. Russ was disappointed at not being able to arrange more than that for me, but I wasn't. Even though I knew Doolittle didn't know me from a hole in the ground, I was ecstatic. He was exactly what I always hoped and prayed he'd be.

Aviation, like show business, had many real stars, but it also had its share of flash-in-the-pan celebrities. A certain amount of ego is essential to any outstanding personality, but sometimes it gets so completely out of hand that the genuine article is overshadowed. The Freeport years of meeting the talented famous had honed my perceptions to a fine degree. This came in handy for screening out the headline seekers from the workers rewarded with headlines. In aviation it came down to those who flew because they loved it, believed in it, and wanted to make it safer for those who'd follow in their footsteps. For more than half a century Doolittle has topped that list, just as I knew he would when he first gave me that forthright handshake and flashing smile, way back in 1929.

I have attended many air meets both before and since the one in Cleveland, but none was more colorful or exciting. The flashing wings of vividly painted aircraft darting about an azure sky gave a fleeting illusion of tropical birds in an aviary — one that was quickly dispelled by the high-pitched whine of supercharged engines and the pungent odor of high-test gas. Offsetting this dazzling speed and color was a solemn daily procession of silvery dirigibles ranging in size from the stubby little Goodyear blimps to the huge, sleek, cigar-shaped *Los Angeles,* whose bulk blotted out the sun when she made her first pass in front of the grandstand.

On the *Los Angeles*'s second pass the navy unveiled its big surprise. A navy fighter plane took off from the field and flew up "into" the dirigible's interior, where it was held by their secret hooking device that deposited it in a "hangar." After a few minutes the airplane came back into view, its propeller spinning lazily as the hook lowered the aircraft. At a given signal the plane's engine was revved up, the hook withdrawn, and a new day dawned for military aerial reconnaissance.

I got so carried away watching these displays that on two occasions they had to page me from the stands to get back up into the air myself. Not that I didn't thoroughly enjoy dropping jumpers and towing Frank around, but I hated to miss out on what everyone else was doing. There's much to be said for drinking in deeply the sights, sounds, and smells of such a colorful mélange. I can go back and enjoy it all over again any time I wish.

By midweek the dust kicked up by tail skids was a serious problem. In fact, when Doug Davis screeched by at 208 mph (the fastest speed then recorded for a civilian aircraft) in a speed-wing Travel Air, I didn't see him at all. The visibility problem was partly to blame for his having missed one pylon, but he went back, circled it, kept on, and won the race!

When the tallies were in, Doug had beaten every military entry in the race. To Walter Beech, the plane's designer-builder, who'd been trying to sell his designs to the government for years, it was sweet victory indeed. When all this hit the papers, the public outcry over military appropriations to build planes that couldn't get within sniffing distance of a civilian aircraft built with private funds put a grin on Walter's face that wouldn't come off.

Also by midweek of the show Deke Lyman had managed to obtain the Department of Commerce report on Red's crack-up. The report claimed that the weakness was not in the ailerons, but in the metal fittings attaching them to the wings. This could mean only one thing — metal fatigue. Was Cessna using old materials?

A sister ship to Red's had been flown in the same race by a Joseph McGrady of Hartford and had almost duplicated the first disaster. Spectators saw McGrady, who was coming in right behind Red, wobbling dangerously as he pulled up from his low pass over the finish line. Examination of his ship after landing revealed that two of three hinges on the left aileron were ripped off, and the rear wing spar was broken at the center of the left aileron. That plane was promptly ordered grounded. Nothing was done, however, about recalling the planes already sold, nor was the Cessna Airplane Company stopped from further production. Bitter in my feelings against these injustices, I rejected the opinions of C.B. and Deke, who insisted that this indisputable proof would surely force the department's hand. At eighteen it's hard to stay cynical, but I managed – every time I remembered that my friend was gone.

Watching Jimmy Doolittle barrel roll out of the top of a loop with the grace of a ballet dancer brought hot tears to my eyes. Red would have loved to watch that ... only now he never would. I pretended the sun was in my eyes and looked away.

I greeted with enthusiasm an invitation from Amelia to attend a meeting of women pilots to be held in her suite. She made a concise presentation of her hopes and aims that an organization to commemorate the Powder Puff Derby and consolidate the gains made in the aviation industry by women would be formed and carried on. Everyone was in complete accord until she brought up the matter of big racing purses in free-for-all competitions. This time I was pleasantly surprised to be joined in my prior objections regarding safety (and for the identical reasons) by Phoebe Omlie and Lady Heath. We therefore did not become charter members of the group, who later called themselves the Ninety-Nines when it was discovered that there were ninety-nine licensed female pilots in the United States. However, it was a friendly disagreement, and we all wished one another well.

The first scheduled pylon race for women in the light plane class was scheduled for that afternoon. Because of other chores, I couldn't compete, but I accompanied Lady Mary onto the field. Before we found out that Phoebe Omlie was going to fly it in her Monocoupe, Lady Mary thought she'd be the only veteran in the race. She decided not to go into it but made the mistake of telling the press that it "wouldn't be fair to the other girls — they're all amateurs, you see."

Understandably this didn't endear her to the

other entrants. I tried to smooth things over by explaining to those I could get to stand still that Lady Mary didn't really mean to be as arrogant as she sounded. She simply had so much more air time in small, light aircraft than the other competitors that there wouldn't be any competition for her. I went back to Lady Mary and suggested she mend her fences immediately — as she did by throwing her arms around Phoebe and saying that now everything was just fine.

Mollified, everyone lined up for takeoff, and Lady Mary promptly shaved the first pylon like an Italian barber. From then on it appeared to be no contest as she breezed home far out in front of the field. But when the figures were in, it turned out that Phoebe in her trusty Monocoupe had actually bested the time of the Great Lakes trainer flown by Lady Mary. However, one of the judges insisted that Phoebe had cut a corner by not properly circling a pylon early in the race and declared Lady Mary the winner. Lady Mary vehemently protested a win on this technicality and insisted she herself had seen Phoebe go back to recircle the missed pylon. This bit of sportsmanship cost her the race, but it also completely dissolved the past resentment held by the other contestants.

I asked her about it that night at dinner. She

had just started to fly for the Great Lakes Company, and losing a race the very first day was hardly the best way to ingratiate oneself with a new employer. She airily brushed aside my concerns as she puffed some of that awful scented Turkish tobacco in my face.

"I could have easily beaten Mrs. Omlie legitimately – in fact, I'm quite sure I did. But I knew from the expressions on the other girls' faces when that pylon business came up that they were about to hate me forever."

"Are you saying that she *didn't* recircle that pylon?"

"I have no idea. That's why I protested so strongly – but as for seeing her do it . . . who looks back during a race, for goodness sakes!"

"Then what you're telling me is that you deliberately threw the race. . . ."

"No, no," she explained patiently. I hated it when she became patient. It almost always meant that she was about to point out something I should have figured out for myself. "By signing on with Great Lakes, I'll be spending more and more time in the United States. In view of that it scarcely makes good sense to start off by making enemies of a group of important American aviatrices, don't you agree?" Her eyebrows arched at me in their most imperious manner.

I nodded and wished she'd thought of that before the race. Oblivious to my silence, she attacked her steak with relish. And after a short pause to marvel at the speed with which she could make a serious error, decide she was wrong, right it, and have everyone eating out of her hand, I did the same.

One thing was for sure: When opportunity knocked on the door, Lady Mary was right there with her hand on the knob.

The day after the women's pylon race I almost didn't recognize my old friend Miles Browning when confronted with a naval officer dazzlingly resplendent in summer whites, replete with the ribbons, gold buttons, and gold-decorated shoulder boards of a lieutenant commander. I'd met Miles when Lady Heath and I were flying in separate planes to the Miami Air Races the previous winter. She had already left Hampton Roads for Langley Field when my Avian's Cirrus engine blew a cylinder on takeoff and I set down in a nearby swamp. Miles, the Hampton Roads base commander, never let me forget that he had led the rescue squad that fished me out of my soggy surroundings.

He explained that he had absolutely no business being in Cleveland now, but the newspaper reports were so exciting that he couldn't bear to miss the whole thing. He then left it up to me to suggest how he could obtain a hotel room, and when that was done, how about an evening on the town, dining and dancing? Miles was the best dancer I've ever met in my life, so I magnanimously gave Deke and C.B. what must have been a welcome evening to themselves and got Bill Lancaster to ferret out a room for him. It was a delightful evening, and I was still on cloud nine the following morning as I hurried down to meet Deke and C.B. and tell them all about it. I found them waiting for me and strangely quiet. Deke said, "I'm afraid we've got bad news, Elinor. Lady Heath just crashed into a building out near that factory with the big smokestacks. They don't think she'll live. . . ."

I stared at him, uncomprehending. First Red . . . now Lady Mary . . . no . . . no . . . *not* Lady Mary. Mutely understanding, they shook their heads. We got into a cab and headed for the hospital, where Bill Lancaster met us at the door. One look at his face was enough. The doctors had told him she would be in surgery for some hours as they probed for bone fragments that were exerting dangerous pressure on her brain. Even if this was successful, they said chances were she'd be blind or mentally impaired. She had sustained a serious skull fracture, broken jawbones, and internal injuries that hadn't been completely assessed yet. I sat down before my legs buckled under me.

Lady Mary and I were not intimate friends — there was too much of an age gap for that — but we were *good* friends, the kind you count on no matter what. It had been that way ever since that first day when she had insisted on taking me to tea at the Garden City Hotel on Long Island.

The Garden City Hotel's heyday had passed with the turn of the century, but Long Islanders loved it for its memories of past Gold Coast glories. I will never forget the baffled expression on the waiter's face as Lady Mary ordered high tea with "fresh strawberries, whipped cream, tea with lemon and cream — and oh, yes, some of those little iced cakes over there —" She waved a hand heavy with jewels. An English high tea probably hadn't been ordered in that hotel for forty years, but the staff rose to the challenge. There was quite a scramble over the strawberries (fresh ones weren't in season), but in due time the hothouse variety and the balance of the repast were served in solemn ceremony to her ladyship and friend.

Lady Mary's admiration for my flying was what started the whole thing off, although it was awhile before she believed my age. When she did, she talked the Canadian Avro Company into hiring me to demonstrate its aircraft to visiting VIPs, her reasoning being that "if a sixteen-year-old could fly its planes, couldn't everyone?"

We flew several long cross-countries together always in separate planes. As I waited anxiously in the hospital lounge, I could hear her during an evening spent at the old Ritz in Washington, urging me to think about the necessity of a rich elderly husband in my future. She was appalled at my naively romantic approach to marriage. As for my American sense of independence, she assured me that it was "all a mirage, my dear – simply a mirage! You must have the kind of man who already has riches and position – the kind who is wealthy enough to assign or hire staff members to handle the details of your career. Look here" – she snapped open a jewel case fit for an Indian rajah – "I suppose you're going to tell me that class ring you're wearing is all you'll ever need."

No, I wasn't going to tell her any such thing, nor was I about to confess it wasn't even mine but belonged to the only boy in high school who ever paid any attention to me. He was now "pinned" to someone else, and when I tried to return the ring, he had refused it. I wore it now so I could look mysterious when curious reporters asked me whose it was.

But she wasn't about to be put off. Snagging a millionaire husband couldn't be left to chance, and she was obviously disappointed in me for failing to recognize the need for a meticulous campaign in that direction. I tried hedging, saying that if God thought I needed a millionaire, I was sure he'd send one around, but she merely regarded me with amused tolerance.

"My dear, do you think Sir James wanted to marry *me?* Of course not! I had to convince him how good it would be for him to have a nurse at his beck and call, and then I had to see to it that he thought the whole thing was his own idea!" So much for blue-blooded romance!

Like most Americans, I considered an authentic British accent the stamp of genteel aristocracy. It never occurred to me that Lady Mary wasn't a blue blood, or at least as wellborn as her titled husband, who, she informed me, controlled all of the coffee exports from South Africa.

A blow-by-blow description of how she had decided on and met Sir James revealed that he was not her first husband. That gentleman had had the temerity to object to her flying. Naturally *he* had been given short shrift, but after the divorce she was penniless, and Great Britain in the twenties required social prestige as well as ability to make it in the flying world. The flying clubs that dominated the civilian flying scene were pretty tightly controlled by the nobility. Lady Mary, aware that this system was too powerful to overthrow, decided on the "if-you-can't-lick-'em-join-'em" strategy.

First she put together a list of the eldest and wealthiest bachelors in the British Empire. She narrowed that down to those needing nursing care (she was a practical nurse) and came up with about five eligibles. The most likely prospect was Sir James Heath of South Africa. Just how she became a member of his household was unclear, but I was so fascinated I didn't want to ask pointed questions.

Predictably she became Lady Heath and, even more predictably, became the first pilot to fly solo from South Africa to England. This flight required all sorts of special dispensations to fly over forbidden areas, and there is no question that Sir James's exalted position opened doors for her. On the other hand, she had become an excellent pilot, and it saddened me that women should be forced to resort to such subterfuge. Still, that's the way it was at that time.

As for Sir James, he seemed happy with the arrangement as long as she dropped in on him a couple of times a year. And in all fairness, she *was* genuinely fond of him. For his part, he never interfered with her flying and was particularly proud of the South African flight. For her the prestige of having a ladyship at her disposal was making her a very wealthy one indeed. Astute with figures, she carefully pyramided her funds, dallying in British and Canadian aircraft stocks as a sideline.

Whenever we traveled together, she always requested a suite with telephones in the bathroom. "One must be in touch with one's broker at all times, m'deah!" My amazement at how the other half lived sent her into gales of laughter.

I thought back to the time we had flown into Chicago on a bitter, wintry day. We were met by a flotilla of photographers and reporters wanting pictures and quotes. I tried to escape to the warmth of the airport office, but Lady Mary was in her glory. Turned out in a stunning leopard-skin coat and beret, she was flattered at a female reporter's compliment and pirouetted around to show it off. "It *is* smart, isn't it? I shot it — in Kenya, y'know." She left the open-mouthed scribe staring after her as she led the way into the hangar.

I would always be in her debt for her support and understanding during my trials with Putnam. His powerful opposition made even my parents have second thoughts. They didn't want me to suffer this kind of harassment, feeling that nothing could possibly be worth it. Lady Mary, on the other hand, was so delighted at my having actually made him bare claws and teeth that her listening ear was always cocked in my direction.

It was her contention that a powerful enemy keeps you on your toes. Well, G.P. had certainly done that for me. But with the support of Lady Mary and other loyal friends, it *had* turned out just as she predicted. And now she was in far more trouble, and I was helpless to lend her the same helping hand. It wasn't right at all, but then, as G. M. Bellanca used to say, "Whoever said life was fair?"

I stared out the hospital window and wondered what lay ahead for Lady Mary. The glory of the South African triumph was far behind now. Bill showed me a cable he'd just received from Sir James saying he "had no plans to come to America at this time." This — after being notified his wife was at death's door.

Stunned and dry-eyed, I listened as Bill related what he knew of the crack-up. She had taken a mechanic up on a promised joy ride and, after a skillful aerobatic display, put the ship into a steep sideslip prior to landing. She must have known about the smokestack — we all had been avoiding it for the past week. Whether her attention wandered or the stack was blotted out by a wing, no one knew, but she struck the stack, bounced off, and hit the roof with terrible impact, bursting through the factory roof to the floor below.

Another version came from Great Lakes executives who said she was practicing dead-stick landings and was without power when a gust of wind blew her into the stack. Luckily the mechanic's injuries were slight, but he was unable to furnish the essential technical details that could reconstruct the true causes of the crash.

With heavy heart I got up to leave. I knew what she would want me to do. The car waited below to take me out to the airport. *That* show had to go on. Extracting a promise from Bill that he'd get word to me the instant the surgeons came out of the operating room, I left Lakeside Hospital.

The crash cast a pall over the air meet that day with both pilots and crowd very subdued. But the next day things gradually returned to normal as Al Williams and Jimmy Doolittle sent the spectators into more spasms of roaring excitement with their display of inverted stunting.

I was ashamed to discover that I, too, momentarily lost sight of my friend's tragic accident as I thrilled to the exploits of the two aces while praying they'd stop before the airplanes came apart.

The word from the hospital was still in-conclusive when I had to leave Cleveland for Indianapolis, the next stop on the tour. I couldn't see her before I left because she was still denied visitors, so I bought her the biggest bouquet of her favorite red roses I could find. Even if she couldn't see them, the nurses would tell her. I wrote a long letter trying to express what was in my heart, knowing it was totally inadequate, but it made me feel better.

Her acknowledgment caught up with me in St. Louis a week later. She was sitting up in bed, her eyes bandaged, dictating her cor-respondence to a secretary, and she apologized for being so tardy in getting back to me!

9

Jumpers and Other Hazards

The Cleveland races ended on September 2, 1929, and our tour for Irvin resumed the next day. I took off in such a flurry of "keep-your-nose-down-on-the-turns" and "be-sure-to-drop-in-on-us-in-Oshkosh-Chicago-St. Louis-Laramie-Pocatello-Phoenix-Clovis-Kingman-Dallas" or wherever, that we were well on our way to Indianapolis before I saw C.B.'s note.

I pulled out the map case and saw that the maps Red had given me had been replaced with a completely new set, all meticulously marked and ruled to take me as far as Wichita. Touched by this thoughtfulness, I turned the maps over and found C.B.'s neatly typed message to "stay away from gliders, stick to parachutes, and don't forget to write your mother!" Well, that was more like it. Our last conversation had been a bit on the down side — at least for me.

C.B. was the only one to whom I confided my dream of becoming America's number one aviatrix — and he wasn't too thrilled with the idea. He credited me with the ability to achieve my goal, but he repeatedly emphasized the long road ahead, insisting that I had little idea of what such success would entail. He reminded me of Katherine Stinson's complete retirement from the public eye in 1922.

C.B. knew that I revered Stinson. Her flying exhibitions had fired my youthful imagination for years. Spanning ten years of aviation's most experimental and dangerous era (1913 through 1922), her brilliant performances had yet to be equaled, and it was questionable if they would ever be surpassed. She had triumphed over the challenges posed by each facet of aeronautics as she broke all existing records for long distance, solo endurance, and speed with almost monotonous regularity. At the same time she managed to pioneer air-mail routes and raise millions of dollars for the government, flying Liberty Bond tours during World War I.

One of her most spectacular aerobatic exhibitions was staged over Lake Michigan on a moonless night when she stunted with lighted flares attached on her wings. Yet she was feminine to the core and wasn't above posing for the press in flying gear with one long curl draped coyly beneath her helmet. But she never traded on her sex, always competed on an equal footing, and at one point bested even the military in long-distance flying.

According to her brother Eddie, Katherine had the publicity instincts of a P. T. Barnum with the moneymaking acumen to match. She was the first American pilot invited (and paid) to appear before the royal families and government heads of Japan, China, the Philippines, Germany, Austria, France, and England.

C.B. felt that Katherine's early retirement was due to the isolation of the life she was leading. Through Eddie, I learned that her retirement had been forced by a severe bout with tuberculosis, but he'd sworn me to secrecy, so I couldn't explain that to C.B.

Actually C.B.'s objections boiled down to wanting me to be available for that great day when some unknown Prince Charming would arrive to carry me away from "all this." Life at the top was bound to be lonely, he warned, and the demands of my occupation would isolate me from romantic contacts. There was no denying that he had a point. My concentration on skill building in order to get this far had already set me apart from the normal socializing of my peers – not that I missed any of it. And looking up at the clear blue dome above, lightly dotted with baby snow white clouds I was about to slice apart with my wings, I wondered why anyone would *want* to be taken away from all this – ever. That night I wired C.B.: "Your maps were great. Hit Indianapolis on the nose, and it didn't hit back. Also wrote my mother. Best – Elinor."

The newsreels of the mass drop in Cleveland preceded us in theaters all over the country, and that, plus the newspaper coverage, drew the enormous crowds that turned out in Indianapolis, St. Louis, and Kansas City. I became adept at landing either ahead of or behind them as I tried to cooperate with local police and airport personnel who did their best to hold them behind the lines. The public still refused to recognize the deadly danger of a spinning propeller, and this was becoming a nightmare to me as the tour progressed.

Wichita was the worst of all. We flew in on a Saturday afternoon to find the roads jammed a couple of miles behind the airport. Peering over my shoulder as I circled around to land, Gene observed, "Either we're the only show in town, or somebody's lifted Prohibition in Kansas!" Well, no one had turned on the liquor tap in that most arid state of the Union. They were, indeed, waiting for us.

Wichita was one of the busiest civilian aircraft manufacturing centers in the United States, so there were plenty of personal friends on hand to greet us. All my pilot friends were curious to know what it was like to take off with a full load of passengers and dump them over the side, so I had a filled cabin on every flight. Interestingly none of these flyboys took advantage of Bert's invitation to join him and Gene when they did their exhibition jumps. To be honest, it made me feel a lot better to know that I wasn't alone in my own misgivings. Not that there weren't still some barnstormers around who both jumped and flew with equal impunity, but these hardy souls were an exclusive minority.

At any rate, my friends' noses were pressed flat against the Bellanca's windows as I circled each set of jumpers on their way down. I'd started this safeguarding practice to keep other planes away from them back in Lowell, Massachusetts, when a curious student pilot came too close to Gene and almost chewed into the chute with his propeller.

On the first day of the show Gene and Bert worked unceasingly with the volunteer jumpers to please the crowd. They finally had to call a halt when they ran out of packed chutes. The next day was more of the same, except that something new was about to be added when Bert told me he was going to have two girls jump on the same flight. We'd never done that before. When he brought them over to meet us, Gene and I were impressed with their beauty and evident expertise. Jean Herrick, a stunning blonde, was introduced as a veteran professional jumper. We assumed that Helen Williams, an equally attractive brunette, had the same background. It was an assumption we would soon bitterly regret.

Just then Walter Beech strolled by. "You going to fill up the sky with jumpers like you did in Cleveland?" he drawled. " 'Cause if you are, I'll put out the word to get all my ships out of the air. I never saw so many jumpers at the same time in my life. I watched you bring this baby down with that tow cable hanging out the back of her — like to scared me to death, you did, that day!" I assured him that he was not alone. I was scared to death, too — only I didn't have a choice.

"Well, while you're in town, how about coming over to the factory? I'd like to show you a *real* airplane — not like this pantywaist thing." He gave a disdainful wave of his hand at the *Happy Landings* insignia on the plane.

To a confirmed Bellancaphile like me, those were fighting words, and I dared him to let me show him what this "pantywaist" was capable of. Walter Beech was a formidable competitor in the civilian aircraft market. His Travel Airs were, and would continue to be, among the finest planes produced in America. But he had scarcely dipped a toe into the large monoplane field. Showing off *Gussie* to him was an opportunity not to be missed.

I had just topped up the gas tanks, and that, along with Bert, Gene, the two girls, Walter, and me, would make up a full payload. Unexpectedly the weather obliged by producing a brisk cross wind, so the stage was now set for a bit of derring-do that no other big monoplane of the day could emulate.

Cross-wind takeoffs in the airplanes of the twenties were attempted only during emergencies and were more often than not disastrous. Air blowing across a takeoff path or a runway could get under your wing and roll you over. And of course, there was always the danger of being blown sideways into an obstruction without your leaving the ground at all. Doing a cross-wind takeoff on purpose was akin to putting your finger in a buzz saw to see how fast you'd lose it — except when you did it in a Bellanca. Her carefully designed wing structure could literally scoop up air from a vacuum.

I chatted gaily with Walter while taxiing out for takeoff. As soon as all into-the-wind traffic subsided, I opened up the throttle and started to roll down the field, pointed directly across the regular takeoff path. Walter's eyes widened as he realized what I was doing, but he gamely said nothing and chomped down hard on his cigar. I held a little extra pressure on the rudder to keep it steady before we lifted off and were airborne. Dipping a wing into the wind, I literally hung her on the prop for a climbing turn that shot Walter's eyebrows up into his hairline. In those days the whole state of Kansas was one vast airport, so he knew I could always get us all down, but he also knew that this climbing facet of the Bellanca's performance was unsurpassed by anything in her horsepower class.

We continued climbing to 4,000 feet for Gene's delayed drop. As soon as he jumped, I let down to 1,800 feet, where Bert said he would let Helen jump first, before Jean. He positioned her at the door, holding onto her chute harness to make sure she would fall clear of Gene. I pulled the nose up at his signal as she jumped out. I had already leveled off and was preparing to circle her when I heard a muffled scream. "Oh my *God!* The chute's not opening! DO something, *please!*"

I swung around and dove down as near to the falling figure as I could. The chute streamed out of the pack, but it shroud lines seemed twisted. As I got closer, I could see that her boot seemed to be caught in them, but I saw no sign of a struggle to get free, which would have been expected, for the chute was completely free of the pack. There was still a good chance to inflate it if she got her foot loose. I followed her as closely as I dared. The awful part was my helplessness. Getting under her might temporarily break her fall if the lines didn't tangle in our propeller, but we would undoubtedly lose a wing, and that would add up to four more fatalities. I could only circle, cutting the motor in and out to let Jean and Bert call out directions to help her. But she never responded. Either she had fainted early on, or she was deliberately collapsing the chute. On the ground Gene duplicated our agony at his own inability to help, but he later reported that she never moved at any time during that terrible descent.

Upon landing, we went immediately to the airport's administration office to file a Department of Commerce report. It was here that we learned that Helen Williams was not unknown to the local piloting fraternity, and some interesting facts started to emerge. She had repeatedly asked to be dropped in the past, but owing to her history of heart trouble and mental depressions that had led to several unsuccessful suicide attempts, she was always turned down.

Because of this, she turned to Jean Herrick, who apparently knew nothing of Helen's physical or mental problems. It is entirely possible that this was true, for neither girl actually lived in Wichita. Helen had come from her home in Oklahoma to visit a married sister. Jean lived in another part of Kansas and was staying with friends in Wichita.

At any rate, when Helen learned that Bert was the man to see about making a jump from *Happy Landings*, she enlisted Jean's aid in getting to see him, for Jean's background in parachuting was impeccable. It wasn't even necessary to see him on the field, for Jean's date for that evening was an airplane parts salesman, who coincidentally was an old friend of Bert's. Everything was worked out smoothly by Helen's being invited to be Bert's date on this pleasant social reunion. Bert agreed to carry them the next day on the basis of Jean's professional credits.

According to Bert, they'd gone to a nearby roadhouse for dinner, drinks, and dancing. It was a pleasant evening that ended early. His friend dropped Bert off first at his hotel, then left with the two girls to escort them home.

I met Helen Williams's father at the funeral home. A gentle man, he assured me that he understood completely that it was an avoidable accident and that none of us should have feelings of guilt. Nonetheless, I did. Telling him how sorry I was seemed totally inadequate.

We looked forward eagerly to the inquest in the hope that more facts would emerge. The chute had already been subjected to the most minute inspection and tests, but nothing that could have caused this serious malfunction was found. There was still the possibility that she had suffered a heart attack or had simply fainted when she jumped out. But it quickly became apparent that we three and the girl's parents were the only ones concerned with this facet of the tragedy.

The county attorney was much more interested in making this case his stepping stone to the elective office of district attorney. As reporters from the big news services around the United States started to arrive, our little investigative hearing took on all the aspects of a criminal trial without jury. There was no question that we had stumbled into territory more hostile than that of any vindictive redneck sheriff.

As the most publicized of the trio I came in for the most savage attack. Despite the fact that it was a matter of record that I wasn't present the night of the party – and didn't, in fact, *ever* party – I soon found myself asked to explain just why I'd been labeled "The Flying Flapper" by the press. Telling the county attorney that I had no control over some unknown rewrite man on an eastern newspaper was met with an unbelieving sneer by the attorney, but the judge ordered the question stricken from the record. At that the attorney hesitated, but then he asked if that label meant that I was a member of the flask-toting, short-skirted generation that frequented roadhouses and speakeasies. And, oh, yes, wasn't I that famous law flouter who had illegally flown under the four East River bridges in New York? I fought for composure as I felt my voice breaking. I would *not* cry before this tormentor, but I didn't know just how long I could hold out if he kept on like this.

Before this verbal assault could continue, the proceedings were halted by the arrival of Walter Beech, who asked to testify in my behalf. He severely rebuked the county attorney for wasting everyone's time with such vicious insinuations, reminding him that I was just eighteen, was one of the most dedicated pilots in the business, and neither smoked, drank, nor partied. Walter finished his testimony with the statement that he'd be glad to hire me any time I chose to leave Irvin or Bellanca. Walter was one of Wichita's most distinguished citizens, and his words carried much weight. My gratitude was heartfelt. No one had asked him to do this, and the very spontaneity of his gesture renewed my faith in human nature.

At the hotel that night the Irvin Company attorney told us that the reason for this line of questioning was to create enough hostility to justify bringing fourth-degree manslaughter charges against Bert. In order to do this legally the county attorney would have to prove that there was so much heavy drinking at the party that night that all three — Helen, Jean, and Bert — were hung-over the next day. He would also have to establish that Bert either had misinformed Helen in the correct method of chute handling or had been negligent in fitting her parachute harness.

To make matters worse, both Jean Herrick and Bert's friend, the airplane parts salesman, had disappeared, and Helen's sister reported that Helen had not come home the night of the party. While we were still reeling from these blows, the papers came out the next day with banner headlines and artists' drawings making that innocent double date appear to be a heavy drinking orgy with sexual overtones. The only bright note was the reestablishment of my own personal and professional reputation, for the reporters at least had the good grace to quote Walter Beech correctly and in full.

Jean Herrick was finally located a few days later, and she testified that there was no liquor on Bert's breath before she and Helen boarded the Bellanca. When asked how she could be sure, she answered simply, "I was looking for it."

That could be taken several ways, but I was glad she said it, for it indicated that she was experienced and knew better than to go leaping out of the first aircraft available to her. Later she told our attorney to anticipate a suit the Williams family was planning against the Irvin Company.

On learning this, our attorney decided that the ship must be moved from the Wichita police jurisdiction. This presented a difficult problem because the ship was currently under guard at the airport. I was still the only one insured to fly it, and now, as a result of the lurid press coverage, no female face in America (with the possible exception of Carry Nation's) was more familiar to a native Kansan than mine.

Through the pilots' grapevine I learned that the police guards might inadvertently be playing into our hands. One of them had been caught with three of our seat cushions under his arm just as he was about to sell them for $25 each as souvenirs. Tex Bohannon, another old friend from my early Roosevelt Field days, was now flying out of Wichita, and he persuaded the airport manager to replace the guards with civilian personnel. He also talked himself into the job of moving the Bellanca into another, more secluded hangar at the far end of the field.

That afternoon, dressed to the teeth in my Sunday best, I arrived at the hangar, pretending to be Tex's "girl friend." Tex had the motor already warmed up and, on the pretext of showing me the airplane, assisted me into the cabin. I was as awkward as any visitor, mincing along in my high-heeled shoes. Tex made a great show of explaining to me how everything in the cabin worked. Then he climbed out and went over to release the rope across the hangar entrance, telling the guards about the arrangement to remove the ship from that hangar to the one on the other side of the field.

It never got there. I simply opened up the

throttle and flew it right out of the hangar door. If it seems strange that I wasn't recognized, I must admit that this always puzzled me, too. But I had long since discovered that for me, the perfect disguise was to dress up in street clothes. People had become so conditioned to seeing me in pants, shirts, goggles, and helmets that they overlooked the obvious. Without all that flight gear, I was just another eighteen-year-old girl who loved pretty clothes.

Once aloft, I faced some unpleasant facts, for I was over unfamiliar territory with no maps. (C.B.'s were good only as far as Wichita.) Tex had told me about a pipeline that ran straight from Wichita to my destination, Ponca City, Oklahoma. Ponca City was a tiny town southeast of Wichita that was far enough over the Kansas line to discourage any official attempts to track me down.

At 1,000 feet, I saw no sign of the pipeline, so I decided to climb up for a better look. I got to 10,000 feet before I saw it, but at that altitude it suddenly sprang into view, pointing southeast as unerringly as a ruled line on a map.

I remember clearly what I was wearing that day: green silk tunic dress with a knife-pleated skirt, a green felt cloche hat trimmed in the same silk as the dress, an oatmeal wool wrap-around coat, beige high-heeled shoes, kid gloves, and matching bag. I loved that outfit. It was very becoming and, as I've said, a perfect disguise. But it was never designed to be worn while flying an unheated cabin airplane at 10,000 feet. Fabric-covered fuselage frames made it impossible to insulate against the cold without adding weight; if you did that, you had to have more horsepower to carry it, and that meant more gas to fuel the horsepower, and bigger tanks meant fewer passengers – and so on. No wonder design engineers were prone to have migraine headaches!

I couldn't wait to tell Lady Heath about all this. We'd had many an argument over her insistence on flying in street clothes, especially when she wanted me to wear the same when I flew her in an open-cockpit Avro Avian in sub-zero weather. I swore she had Eskimo blood but refused to get double pneumonia just to avoid her ire.

Flying in a separate aircraft, I was bundled to the teeth in fur-lined flight suit, boots, gloves, and helmet. When we landed, she faced the news cameras as chicly turned out as a *Vogue* magazine cover, while I trundled along in her wake looking like something out of *Popular Mechanics*. But I didn't care. I was *warm*.

Shaking with cold, I guided the ship down to a more comfortable altitude, zeroed in on the pipeline again, and was soon over Ponca City. *Airport* was not the word I'd use to describe the landing area below. In fact, I "dragged" it twice, hoping I was mistaken. Tex's hurried description of "two hangars at one end with a big, new windsock" didn't fit this deserted backwater at all. But when I saw the ragged orange cone hanging limply against its bent pole support, I knew this had to be the place, although Tex obviously hadn't visited it lately. I landed, taxied over to the forlorn, weather-beaten shacks that passed for hangars, and shut off the motor.

Anxious to stretch my legs, I got out and walked around. It was even more discouraging on the ground than it looked from the air. There were two windows in each building, but they were too high up for me to get a glimpse of what might be inside. I got back into the Bellanca, found some parachute covers in the back of the cabin, wrapped them around me, and for the first time squarely faced the vulnerability of my position.

I had no luggage with me and little cash. I had just enough gas left to make Oklahoma City in the morning if the weather held. If it didn't, the anticipation of a prolonged stay in

Ponca City was anything but joyful. I was ravenously hungry, not having eaten since breakfast. The cold was seeping through the chute covers, and I had a splitting headache. Maybe if I tried to sleep...

I woke with a start as a flashlight beam played on my face, and I was noisily greeted by Tex and Gene. I was never so glad to see anyone in my whole life and surprised us all by bursting into tears! It seemed that shortly after I left, the Irvin attorney decided it would be best for Gene and me to stay out of Kansas, so Gene had brought all our luggage. In order to get away safely, they had had to wait until dark, a hazard in itself.

Night cross-country flying in the twenties was strictly hit-and-miss, with the accent on the miss. Air-mail beacons were just starting to make an appearance and, in fact, had just been installed on the Wichita to Oklahoma City run. But as Tex pointed out, Ponca City wasn't exactly a high priority on the government's list, so they watched the beacon rather wistfully as it blinked on and off to the west of them until they lost sight of it altogether.

Gene's hilarious description of picking up on automobile headlights and reciting his rosary, while Tex made swooping passes at filling station signs trying to find out where they were, soon had us all in high spirits. In short order, we were composing a letter to our Irvin bosses in New York, who were doing business as usual in the dignified splendor of their Fifth Avenue offices, while we — to quote Gene — "flew our tails off, chasing around the Southwest just two laps ahead of the sheriff!" He had just gotten around to suggesting we ask the company for combat pay when the full realization of how hard they both were working to keep my mind off some pretty depressing facts hit me. Here we were, two men and a girl, shivering in an unheated airplane cabin that was getting colder by the minute, stranded in this Godforsaken place at least until morning. Tex's thermos coffee had long since run dry, and we were hungry enough to eat the seats, but what, if anything, could we do about it?

I never found out, for we were rescued from this bleak situation when the beams of two automobile headlights pierced the gloom of the cabin. "Hi there — need any help?" someone called out. *Did* we! The three of us got jammed into the cabin's doorway together in our hurry to latch onto this good Samaritan before he got away.

Our rescuer turned out to be a newly rich Indian, complete with braids, black hat, and feather, who'd been too smart to sign away his land rights to a rapacious oil company. His Packard sports car was a block long and trimmed in enough chrome to cause a shortage of the stuff in Detroit.

He took us to a nearby roadhouse for dinner. I'd never been to one before, and in view of the role that a roadhouse had just played in Wichita, I'd have been happier to go almost anywhere else. But we all were famished, and the food was delicious. Besides, it didn't look like a den of iniquity, although there were setups (bottles of ginger ale and soda on each table). I saw some guests pouring what looked like liquor into their glasses from flasks, but that was all. No drunkenness, no wild behavior. Either Wichita's reporters had livelier imaginations, or I was in the wrong place.

Our host (for he insisted that's what he was) was an eastern college graduate, despite his picturesque attire. Having handed his hat and what looked like a Mexican serape to the hat-check girl, he summoned the *maître d'*, whose obsequious bowings and scrapings soon made it obvious that our patron was more than a well-heeled Indian with a college degree. As it turned out, he owned the roadhouse, the hotel he put us up at, and quite a lot of Ponca City. He also had two Stearmans and a clipped-wing Waco to fly around in and was one of the best aerobatic pilots in that part of the country. But he got his jollies by hanging around the one barbershop in Ponca City and letting all and sundry take his picture to show the folks back home what a *real* Indian looked like.

I never learned just how much of a put-on it all was. We didn't stay in Ponca City that long, and I don't even remember our host's name. But if this was a sample of the white man's treatment by the American Indian, Gene and I were willing to take on General Custer all over again.

We got word from Wichita that the charges against Bert were about to be dropped for lack of evidence (and no further proof ever did surface regarding the cause of the Williams tragedy). Unfortunately the delays caused by the county attorney's ambitions had now completely disrupted our tour schedule. The other cities on our route were predictably lukewarm about carrying out their commitments to the Irvin Company. By the time we flew into San Antonio it was obvious our triumvirate was about to be disbanded. The company owned the Bellanca and planned to use it for the purpose for which it was designed – to train jumpers.

Good-byes are difficult under the best of circumstances, which these certainly weren't. The events of the past few weeks had caused Bert to become withdrawn, but Gene was his usual ebullient self. At the last minute, though, both Gene and I cried a little. What is it with the Irish and the Italians?

I flew back to New York via commercial transport. It was still the day of cold box lunches, topped with an apple and/or a dried-out cookie. I took one look and closed the lid. Better to skip lunch altogether and devote time to the spectacular view of gauzy mares' tails sweeping across the aquamarine sky. Normally I would have been lost in joyful contemplation of such a display of nature's artistry. But today my thoughts were demandingly intrusive. The tragedies of the past month defied my understanding.

None of them, with the possible exception of Lady Heath's, could have been avoided — and all reports on that crack-up weren't in yet. Red's Cessna certainly hadn't given any signs of structural weakness on landings prior to Boston. I still don't know what happened with the one I'd flown up in Connecticut, but in any case I had stayed out of Cessnas ever since.

The cause of the Williams disaster was unclear. But the fact that she was a novice was far from being the whole answer. How many hundreds of jumpers had we dropped on the tour who were admittedly making their first descent?

The "you'll-go-only-when-your-number-is-up" philosophy had always annoyed me, indicating as it did that we all were slated to crash some time. Convinced that carelessness was the basic cause of most aircraft accidents, I had always insisted that sensible observation of the safety precautions made flying safer than driving your car. Remembering that I'd picked up a lot of this from Red caused the tears to well up again. But how was I to deal with this mysterious unknown factor that had appeared with such stunning force three separate times in the past month?

Making lemonade out of lemons was a daily occurrence in a pilot's life. Perfect flying weather could turn foggy on a moment's notice. Engines still cut out for mysterious reasons, forcing you to land immediately, sometimes on terrain as hostile as Mars. But somehow you always had a fighting chance. There seemed to be no such chance in this new dimension. It was something you couldn't anticipate – or avoid.

My confusion must have been obvious, for a sympathetic lady passenger asked me if I was ill. I forced a smile and shook my head. If only it had been that simple!

I was pleasantly diverted from these gloomy thoughts when the copilot of the big Fokker turned out to be an acquaintance from the old Waco days. We exchanged news of old friends, and of course, he wanted to know all about my recent troubles. He tried to lighten my mood by inviting me up into the cockpit to meet the captain and take a turn at the controls. This piloting courtesy was a common practice at the time. Unfortunately I was traveling incognito (in street clothes again), and I couldn't help overhearing a passenger seated nearby say bitingly to her neighbor, "Wouldn't you think they'd be more discreet! She's a bold one, all right. What do you bet those men are married?"

Ben heard her, too, and after he helped me into his seat and introduced me to the captain, he went back down to the indignant protester and said loyally, "Lady, when you can fly as well as that girl, we'll be happy to take your application for a pilot's job. She's probably got more hours in the air than the captain and me put together. The airline would love to have her fly for them — but they can't afford her."

The door to the cabin was closed when Ben made these chivalrous pronouncements. But when I came back to my seat, it was obvious that my status had been upped considerably.

The disgruntled lady apologized and told me what he'd said. From then on until we landed, she and the rest of the passengers kept me busy autographing napkins, lunch box cartons — even some shopping bags. My little session at the controls had effectively dispelled my gloom, and I found the attentions of my newly found friends a welcome distraction.

10

Go West, Young Woman, Go West

When I arrived home from Texas, I asked Mother to turn off the phone and let me sleep for a week. I was emotionally and physically exhausted, and it was some time before I realized that the first part of my request was unnecessary. The phone wasn't ringing – or when it did, it wasn't for me.

Going over to the field to fly Father's latest Waco did little to raise my spirits. If anything, they were lowered a notch by the distinct chill emanating from the Curtiss side of the field and from some of the press. But loyal Herb McCory enveloped me in his familiar genial bear hug as I climbed out of the Waco's cockpit.

"Boy, it's good to have you home!" he boomed.

"Really?" I asked, "I'm glad to hear it because there seems to be a difference of opinion about that from some others around here." In answer to his baffled look, I told him what I had encountered. He harrumphed nervously a few times, saying it was my imagination, but ended by saying he'd look into it anyhow. I watched his broad back retreating across the field and thanked my lucky stars for the unquestioning loyalty of such a good friend.

By the time I got the Waco bedded down for the night he was back with the full story. Eyes blazing, he said, "Well, Sis, you were right. There's a big freeze on. Not that *you've* done anything wrong. After that blast of Walter Beech's in all the papers, they wouldn't dare say anything about your personal conduct on the tour, or your piloting either for that matter. It's just that this is the first time that bunch has really had a chance to get their hooks into you."

"But why? What have I done to make them want to?"

"Nothing but make Casey Jones madder at you than ever, and his pilots know it's healthy to get in line behind him. Also, I'm sorry to say, he's convinced some of my guys to keep their distance by implying that you're not exactly an asset to the aviation world at this point."

"Because of Wichita?" I asked incredulously. Tragic as that accident was, I didn't see how it could be laid on my doorstep and said so.

"I couldn't agree with you more," said Mac, "but I'm just telling it like it is. Jones has never forgiven you for all the publicity and world's records of the past year, so this temporary setback has given him the ammunition he wants."

"But how about Chamberlin, Byrd, Acosta, and all the rest? Does he resent them, too, or is it just me?"

"Think back," said Mac patiently. "Remember when Curtiss wanted to hire you and he talked them out of it? Well, the Curtiss brass has never let him forget it. Every time one of his girl students washes out, they razz him to the heavens."

"But, Mac," I protested, "you remember — he used to be Pop's friend."

"Sure he was, as long as your father was paying for Curtiss instruction. Once Tom bought a Waco and hired Red to fly it, he was no longer putting money in the Curtiss pocket. Don't forget, I thought Casey was my friend, too, until I found out that he was trying to get me thrown off the field after you flew under the bridges. If you'd been a Curtiss student, he'd have thrown every bit of Curtiss clout behind you, and there would have been no problems with the city or the Department of Commerce. Instead, he made it as tough for you and everyone who publicized you as he could. Of course, you have to remember that he's been flying for years and never once hit the papers like that. Nor has he made the money out of it that you've racked up already, and you're still only eighteen years old."

"Well, why doesn't he get mad at Lindbergh then? He's still in his early twenties and he's making more money than I'll ever see!"

"Simple," said Mac patiently. "Lindbergh's too big for him to tackle, and you're not."

I mulled this over and said, "Sooooo . . . if I don't pull out of this slump he's created for me, he'll be on hand with the I-told-you-so's, and if I do come out of it, it will just go to show that a bad penny always turns up, right? No matter what I do, he's got me in a no-win box."

"That's it, kid. He's got that picture on the wall with your name under it."

"Hmmm . . . well, what can't be cured must be endured — or something like that," I said thoughtfully.

"Got any ideas?" asked Mac hopefully.

"None that I'd care to discuss when you've

got that reporter's gleam in your eye. Anyway, I'm currently wavering between becoming a cloistered nun and setting fire to the Curtiss hangars."

"I wouldn't put either one of them past you," he declared laughingly. "But whichever you decide, can I have first dibs on the pictures?"

Back home that night, Mother told me that a Mr. Sarnoff had been trying to reach me all afternoon. "You mean David Sarnoff of the Radio Corporation of America?" I asked incredulously.

"I guess that's who it was," she said calmly. "I didn't recognize the name, so I didn't ask."

I returned the call the next day and was asked to attend an executive meeting in Mr. Sarnoff's RCA office to discuss "something we think will be of interest to you." Whatever they had in mind would be an improvement, I thought wryly. The way things were going, I'd started picturing myself back on Cross Bay, flying passengers off the beach again.

I caught an early train into the city the next morning and found myself contemplating this new opportunity with mixed feelings. What could a huge corporation like RCA possibly want with me? I longed for Lady Mary. She'd get the whole thing into focus in two crisp sentences. I recalled the time she had told me about facing a top-level meeting of Canadian aircraft officials. "I dressed for it to the very nines, m'dear. Furs, jewels – everything. After all, to men like that, big business is a game in which they hold the chips. My being smartly turned out like that made them stop and think that maybe they didn't hold them *all* – and while they were thinking . . ." Her triumphant chortle indicated that the Canadians didn't, in the final analysis, hold all the chips.

I looked down at my demure brown silk with distaste. I had chosen it that morning because it seemed businesslike. And it was – if you carried a pad and pencil and took notes for the boss. I went straight from Penn Station to B. Altman's and purchased an immensely becoming (and horrendously expensive) raspberry wool suit, matching blouse, hat, and gloves. Then I sallied forth to do battle. Lady Mary's flesh might be temporarily weak, but her words were coming through loud and clear!

The meeting in Mr. Sarnoff's office was impressively staffed by a select group of executives in radio and electronics whose names and faces I knew only from the newspapers. Along with Mr. Sarnoff, then executive vice-president of Radio Corporation of America, were Merlin Aylesworth, president of the

National Broadcasting Company, and Owen Young, president of General Electric. I saw just one familiar face in the distinguished assemblage. It was Jap Gude, a Tydol gasoline official with whom I had done business prior to the last endurance flight. His presence alerted me that there must be some sponsors as well as advertising men in attendance. This made me more curious than ever about the purpose of the gathering.

I didn't have long to wait. Mr. Aylesworth opened the meeting and quickly proceeded to outline some plans for network expansion, which included the purchase of an executive aircraft to be used and owned jointly by NBC and RCA. There was skepticism on some faces when he announced that I had been selected to pilot this craft, "provided she has no other commitments, of course." He then pointed out that I had just returned from a 4,000-mile tour as executive pilot for the Irvin Company. I had been asked to come in today in view of Irvin's hearty recommendation. (*That* was interesting. I hadn't heard jot or tittle from them myself.)

Mr. Aylesworth felt that having a female pilot would guarantee them publicity in the venture, to say nothing of promoting an anti-discrimination image. And with that, he turned the meeting over to me!

I still hadn't recovered from the shock of finding out I was even in the running for this high-level post. Nor did I have any idea what he wanted me to say, so I did the only thing I could think of and asked for questions from the floor.

Some of the ad men were extremely knowledgeable about planes and flying, and this helped a great deal, especially when they asked me to recommend an airplane. I was off and running with an enthusiastic recital of the safety statistics of my beloved Bellanca. I was careful to point out how much more comfortably it could be appointed for passenger use than any other monoplanes of similar size. A few of the N. W. Ayer men spoke up for Lockheed, but my emphasis on safety overrode their desire for speed. It was agreed that I would arrange for a Bellanca demonstration, and the meeting ended.

I floated out into the hall, went down in the elevator to the street, and never touched the ground once. I kept pinching myself to make sure I wasn't dreaming. I raced to the nearest pay phone to call G.M. down in New Castle. Disappointingly he was out of town. I could have easily arranged for the flight demonstration through George Haldeman, but this news was so good I wanted to break it to the boss myself. I hung up and called Deke at the *New York Times*. He was as elated as I was, and by the time we finished talking we had convinced ourselves that aviation's bigtime had truly embraced me.

This euphoria was short-lived. Several days later I received a phone call from Mr. Sarnoff's office, saying everyone was so very sorry, but the project had been temporarily shelved as a result of financing problems. They sincerely regretted taking my time, etc. . . .

With sinking heart, I put down the phone. There was something about it that was depressingly familiar. Oh, no – not G. P. Putnam *again!* I called C.B. at the *World* and asked him what he thought. The only proof I had this time was my Irish intuition, and C.B. considered that far from conclusive. But he said he'd look into it anyway, just to humor me – and show me how silly I was. He called back in less than an hour to tell me how much he hated to admit I was right.

Our boy had indeed been in to see Mr. Sarnoff and had suavely convinced him that my name was now so sullied by the Wichita incident that it would be most unwise to employ me. Meanwhile, how about Amelia for the job? She was familiar with the Lockheed, having flown it in the derby, etc . . . One of the Sarnoff staff pointed out Amelia's flying record there, and they politely declined Putnam's offer for her services. C.B.'s informant said that Putnam became livid with rage at this and rudely slammed out of the office.

"Sounds like him," I said dryly. "But why is he so upset? Even if they refuse to hire Amelia, they aren't hiring me either."

C.B. agreed, but his reporter's instincts were fired up. He wanted to know how Putnam had found out so quickly that RCA was even considering hiring me. I hadn't a clue because there had been so many people at the meeting I didn't know. At any rate, the fat was in the fire now.

Just as I hung up the phone, Mother came into the room waving a telegram. It was from Bobbie Trout. She had obtained equipment and financial backing for the first refueling-in-the-air endurance record for women. Would I care to join her?

We had discussed this briefly at Cleveland, so it wasn't a complete surprise. In my present frame of mind, it took me just long enough to back the car out of the garage to decide this was a heaven-sent opportunity to shake the New York dust from my feet. I raced downtown to the Western Union office to wire my acceptance the same day.

At this low point in my life, had I been left to my own devices, my instincts would have led me to fold my tent like an Arab and steal away to the West Coast. But my loyal friends McCory, Allen, and Lyman provided a send-off that would have done a movie star proud. It was complete right down to Penn Station's much-worn red carpet and back-of-the-railroad-car pictures, and I was photographed in my new raspberry-colored suit, carrying an enormous bouquet of matching roses wired by Miles, who was unable to come to New York.

Flying to California in 1929 was not exactly a pleasure jaunt. Despite being heavily publicized as "the best and fastest way to travel," it was truly an exercise in discipline and endurance. Set up to save time, it actually cut little from the excellent train service already available. But the novelty of flying by day, and riding the rails or stopping overnight to sleep, was an experience of sophisticated journeying not to be missed. I boarded the Pennsylvania Limited in New York and went by sleeper to Columbus, Ohio. From there I picked up the TAT (Transcontinental Air Transport) all-metal Ford Tri-Motor for a long, noisy haul to Waynoka, Oklahoma. At Waynoka, I disembarked, got on another train, and woke to greet the dawn in Clovis, New Mexico. From Clovis we droned on (at all of 75 mph) to Los Angeles. The faithful Tin Goose (Ford Tri-Motor) did its best, but even with its nose down on the turns it could go no faster.

Tired, dusty, and deaf, I was warmly greeted by Bobbie at Grand Central Airport in Glendale. What I wanted most in the world was a hot bath and a cool bed, but I started to revive as Bobbie and I got into a nonstop dialogue, catching up on old news and new gossip. Always an avid movie fan, I wanted to know all about the studios, stars' homes, Brown Derby restaurant, and all the rest of Louella Parsons's world. Bobbie laughed and assured me that we would get to it all in due time, but her mother was waiting supper for us, and in the Trout household the film industry's position was invariably below the salt.

Mrs. Trout had prepared a feast that had me worrying about my weight before I got past the entrée. Bobbie's mother was a gray-haired motherly lady of whom I became very fond. She knew little of aviation and seemed slightly bewildered by having a daughter whose interests were centered on planes and flying. A genuine homebody, she was a charter member of that vanishing breed whose concerns are limited solely to their domestic domain.

I got my first taste of Southern California's fickle weather the next day on the drive to the airport. After starting out under a blue cloudless sky, we were suddenly enveloped in a fog so thick we had to pull off the road to wait for the sun to burn through it. Bobbie casually shrugged it off, saying that it happened all the time when heavy cloud banks rolled in off the Pacific, blanketing the countryside. I thought back to Red's warning about night fogs and wondered what our chances would be if the ground were blotted out during a time when we might be forced down. There was evidently no way to plan a flight around these happenings. Such fogs were much too unpredictable. Even twenty-four-hour weather predictions were unreliable — and Bobbie and I were planning to stay in the air for at least a week.

When the fog finally lifted, I was impressed by the abundance of unfamiliar greenery dominated by huge palm trees lining the roadway. With the blue Pacific glittering peacefully in the distance, I had a sudden sensation of having been set down on another planet. But our arrival at Metropolitan Airport in Van Nuys, with the familiar sights, sounds, and smells of a busy airport, dispelled the momentary alien sensation, and I began feeling right at home.

We had just entered the hangar when the

noisy blasting of an automobile horn and the high-pitched squeal of protesting brakes heralded the arrival of Roscoe Turner, an old friend from Roosevelt Field. As flamboyant a pilot as ever graced a cockpit, Roscoe ranked with the best. He was also one of the top stunt pilots in the movie world, and was currently working on a movie for Howard Hughes.

Roscoe believed firmly in the value of self-advertising, and no one within a five-mile radius was ever unaware of his presence. If he wasn't roaring in over your head in an ear-shattering power dive, his personal attire was guaranteed to rivet audience attention and stop traffic in the middle of Sunset Boulevard. To-day he was tastefully turned out in a sky blue tunic and matching overseas cap, each heavily embroidered with gold wings in varying sizes. Creamy whipcord trousers, enhanced by a red stripe edged in gold that started under his arm-pits and continued down to boot tops, drew attention to the highly polished riding boots and snappy Sam Browne belt that lent a military flair to his ensemble. His meticulously waxed mustache, deep tan, and flashing white teeth provided the correct finishing touches to this vision of male comeliness. But just in case anybody came in late, he had a pet lion cub tethered to the end of a gold rope.

Roscoe's arrival just happened to coincide with that of the reporters and photographers who'd come out to Metropolitan to record for posterity the formal meeting of the two former flying opponents who were now teaming up to tackle the most difficult flight ever attempted by a pair of female pilots.

But after one look at the lion, Roscoe, and Roscoe's companions, the newsmen forgot all about us. For Roscoe had brought along none other than the legendary stunt pilots Frank Clarke and Roy Wilson, aviators I'd been hearing about for the most of my flying life. To be absolutely truthful, however, it was the lion cub that stole the show.

Roscoe had been a lion tamer in his youth, so he obligingly put the cub through his paces. He was a fascinating animal, though not too reliable in the house training department. He was so beautiful that I was strongly inclined to accept Roscoe's invitation to "Go ahead — pet him. He loves it." Up to then our feline friend was purring like an outboard motor. But at the first step I took in his direction, the ground shook from a rumble that started deep in his throat. I'd been brought up with cats, but I quickly realized that this was no friendly tabby. In fact, if my ESP was working at all, this golden-eyed beauty with twitching tail was saying "Watch it, kid! The only one around here who scratches *my* ears is that guy over there in the red stripes!"

Frank Clarke was the only pilot I'd ever heard of who'd made a successful midair rescue, and I'd been anxious to learn the details ever since the Wichita incident. It was particularly important to me now. Was there anything I could have done and didn't to help Helen Williams? If anyone could tell me, I was sure it was Clarke, though I didn't tell him why.

He was most obliging and launched into his tale in a manner that suggested it was all in a day's work and he really didn't see why anyone thought it out of the ordinary. He had been flying a camera ship, close alongside another biplane. The photographer in Clarke's plane was grinding away, recording the antics of wingwalker Al Wilson as he cavorted around on the top wing of the ship alongside without parachute or toe holds. Suddenly Wilson stumbled and fell off into space. Clarke dove under him and skillfully caught him on his own ship's top center section, the most heavily braced part of the wing. Gunning the motor to blow Wilson off again while simultaneously pulling up into a climb, Clarke slid the nearly unconscious man into the front cockpit "just as neat as making a cue ball shot in the side pocket!"

We all roared with laughter, but there was a lot of awe in our mirth. I was filled with respect for his ability, but also deeply relieved, for he had made it clear that there was no comparison between the ships we'd been flying on those occasions. My Bellanca did not have an exposed cockpit, and her whole outer surface was as smooth as glass. Her center-section bracing would have been no help at all because a high-winged monoplane presented nothing for Helen Williams to grab onto, had she been consious at the time — and I always felt she wasn't. But aside from all that, her trailing chute lines would have inevitably tangled with either the propeller or the tail section. Clarke probably wondered why I insisted on shaking his hand that day, but it wasn't the time or place to explain.

Up to then I had always been inclined to doubt a widely circulated story about Clarke's winning back a reluctant ladylove by running the wheels of his airplane up the side of her hotel in order to toss a rock-laden message into her bedroom window. But today's meeting changed all that. From now on I was a true believer where Frank Clark was concerned.

Roscoe and Frank were teaching their boss, Howard Hughes, to fly, although they were always careful to "keep him on a short leash, so he can't fire us if he gets some other nuts to do our jobs cheaper." In view of the skills required for aerial films, the pay scale was outrageously low, and Hughes, like every other producer in Hollywood, was not averse to taking advantage of the going rate.

Ultimately they made a fine pilot out of him, but at the time they were teaching him just enough at each session to whet his appetite for more. This technique backfired just once, when Hughes attempted to put one over on his instructors by taking off in a Thomas Morse Scout with a rotary Gnome engine. The TM was a World War I airplane that should have been flown only by an expert. Its revolving engine gave it so much right-hand torque (pull) that one never made a right turn with it under a couple of thousand feet.

Hughes didn't know this, so when he made what should have been a graceful right turn into the wind, the ship promptly dove into the ground. Amazingly enough, he was not seriously hurt, causing Roscoe to later remark, "You know, it made Frank and me feel real bad. He wasn't the most likable guy in the world, but that's a helluva way to lose your meal ticket, no matter how you look at it!" Roscoe wasn't being callous. He was telling it like it was. Hughes had not been ready to fly the TM, and his arrogance had almost cost him his life.

The ridiculous wages paid by the studios finally led Pancho Barnes to form the first protective organization for movie stunt pilots, an important step toward correcting the situation. Florence ("Pancho") Lowe Barnes was a descendant of one of California's most prestigious families, the Thaddeus Lowes of Pasadena, and an excellent pilot. Unlike Highes, she used her social clout to curb the scandalous exploitation of professional fliers by the movie industry. She ran a tight ship and was respected by both pilots and producers. Ultimately this group was known as the Motion Picture Pilots' Association (MPPA) and became one of the most powerful organizations in the film industry.

I didn't know Pancho well. I had met her briefly in Cleveland and was impressed with her western breeziness and her skill in handling her airplane, but I was soon to learn that Florence Lowe Barnes was an individualist all the way. Her casual dress, salty conversation, and predilection for foul-smelling cigars, which she lighted with kitchen matches struck on the seat of her pants, did not project the image of a wealthy patron of the arts her birthright obviously entitled her to.

Pancho didn't aspire to be a dedicated pilot. She *was* one, and because of her love of aeronautics and her skill, doors that no other wealthy sportsman-dilettante could possibly have entered were opened to her. I once asked her how she had acquired the name of Pancho. She chewed reflectively on her cigar and said it had happened so long ago she'd almost forgotten. In her youth she'd eloped to Mexico with a minister named Barnes. The marriage was short-lived, but her friends likened her rebelliousness to that of the Mexican leader Pancho Villa. For the rest of her days she remained Pancho, and it suited her very well. It was certainly a great improvement on her given name of Florence.

Pancho owned a string of thoroughbred quarter horses that she kept south of the border, so she did a lot of flying in and out of Mexico, basing her own ship at Metropolitan Airport. She flew a brand-new Travel Air Mystery ship, one of the few that were privately owned. Her love of flying and sincerity of purpose were as much a part of her as the ever-present cloud of cigar smoke, and one had to like and admire her. But it turned out that our friendship was to be confined to the airport and its environs. Despite a reputation for being the most lavish of hostesses, she bluntly told me that I was too young for the goings-on in her home.

I can't say that I took this news either meekly or in good grace. I'd been told that she stocked the best bar in that part of California, and this would, I knew, attract some of aviation's most dedicated imbibers. Not being a drinker myself, I knew I was missing out on some of the most fascinating hangar flying stories around, and I felt very sorry for myself. Years later, while dining with Richard Halliburton, author of an enormously successful string of adventure books, he touched lightly upon the subject of some of Pancho's parties he'd attended. Only then did I realize how wise Pancho was and how out of place I would have been. In her later years she was persona non grata at Edwards Air Force Base for the goings-on at her "Happy Bottom Riding Club."

I didn't get to fly the Sunbeam, that first day on the field, but what I saw of her I liked. It was a large biplane powered with a 300 h.p. J-6 Wright engine, and its ample cabin, set in front of the pilot's cockpit, was jammed full of gasoline tanks, which, with blankets thrown over them, would be our beds for the rotating rest periods Bobbie and I planned on. I did notice that her landing gear seemed a little narrow-tracked, but her weight and oleo strut landing gear would probably take care of any tendencies toward swinging out of control.

I was more interested in getting to know our refueling team, Paul Whittier and Pete Reinhardt. I liked both men on sight. Their sincerity and hope for our project were unmistakable. But I was not so enthusiastic about their equipment because our ships were ill-matched for the job we proposed to do. The refueling craft was a Curtiss Pigeon, one of the first air-mail carriers to be used by National Air Transport back in 1925. Powered with an ancient Liberty engine, she was the only ship available with enough fuselage area to carry the gas load we needed to keep the Sunbeam in the air. Built for reliability rather than speed, the Pigeon had performed admirably in her day. In point of fact, she still did, except for the disconcerting fact that her engine kept breaking down. Parts for it were scarce and difficult to come by. The financing for the flight did not cover a new engine for the Pigeon. Even if another Liberty engine had been obtainable, it is questionable if it would have been in any better shape than the one we had.

The plan was that Paul, a wealthy young sportsman pilot, would fly the Pigeon, and

Pete, a former barnstormer and stunt man wing walker, would man the gas hoses while lying on his stomach and hanging out of a hole cut in the bottom of the Pigeon's fuselage. Because of my pylon racing experience and familiarity with aerobatics, it was decided that I would do the contact flying and Bobbie would handle the hoses. After a few practice flights I knew this operation would be no snap. For one thing, the cockpit of the Pigeon was so high up in front that Paul couldn't see us when we were hooked onto the hose. For another, Pete could look directly down on us but couldn't communicate with Paul. This left it up to me to position the Sunbeam so that Bobbie could latch onto the writhing hose, and I had to then keep the engine throttled back far enough to stay in position without stalling off into a spin. Since Paul couldn't signal me when the ailing Liberty started to sputter, I soon became adept at ducking out from under her landing gear when she would start sinking down on us. Lumbering and cumbersome under normal conditions, under stress she had all the glide of a drag anchor. Paul gamely wrestled her down in so many forced landings that Pete said they made more touchdowns than an overweight sea gull.

On the ground and on paper, the Sunbeam appeared to be just what we needed, but my first flight in her convinced me this optimism was misplaced. In the air she was about as stable as a roller coaster with a loose cable. My first landing almost became a disaster when she whipped around in a vicious ground loop that took me totally by surprise since I had stopped rolling at the time. Even Bobbie's assurances that the same thing had happened to her the first time out did little to restore my original confidence. I knew we were in for big trouble with an airplane that had to be flown every single minute with all the concentration of a test flight. Still, it was all we had to go with, and S. George Ullman, our business manager, had made commitments we had to live up to.

The bright spot in this gloomy scene was Bobbie's flawless performance. She was fearless about hanging onto that hose. As soon as I maneuvered us in close enough, she never missed catching it even once. Although she knew full well that each time we connected she faced the possibility of getting drenched in high-test gas, she never flinched. A fitted leather harness held her in the cockpit when she stood up, but there was the ever-present danger of her being hauled out in midair if the harness broke during some unexpected air turbulence. There was no possibility of her wearing a parachute. It was too cumbersome and would be too heavy for her to work in.

We'd been up about twelve hours on our first attempt at a record when I jockeyed the ship into position to take on some gas. We were out over the water near Catalina when some rising heat buffeted us about sharply, and the Pigeon dropped so fast I had to dive quickly to avoid having her wheels crunch into our top wing. This broke the contact, and Bobbie was soaked in high-test gas from head to toe when the hose was yanked out of her hands. Pete cut himself deeply in trying to keep the lines straight, and I could see the blood streaming up his arms in the prop's blast as he fought to hang onto the hose.

I swung around and headed back immediately to Metropolitan. We all were greatly relieved to find that Bobbie wasn't blistered or burned. Nor were Pete's cuts as bad as they'd first appeared. But the whole incident gave me pause. Most of our troubles stemmed from our inability to synchronize the speed of both ships. If I kept the Sunbeam at cruising level, I would be flying ahead of the Pigeon, towing her through the air. We had been very lucky so far, but I didn't want to push that to its limit.

Searching for ways to improve or change our refueling techniques, I thought back to the first week of this record attempt, when three of the navy's High Hat Squadron had flown up from San Diego at Admiral Mofett's request to make sure I knew all the tricks involved in such close contact flying. According to them, I'd picked up the procedures in record time, and they'd even made a little ceremony out of pinning a cardboard A+ on my blouse.

Still, while I was working with them, the Sunbeam wasn't heavily loaded, so maybe the trouble *was* with me. Was I overcontrolling? But if I was, how had I managed to get out from under the Pigeon all those times her motor let go? The only way I could tell for sure if I was doing something wrong — and to find out how to correct it — was through aerial photography. But I couldn't very well ask the newsreel cameramen. In those days, you didn't dare own up publicly to any doubts. We "daring and intrepid airwomen" (as reporters invariably called us) weren't allowed to have a care in the world. Spoiled the image.

Maybe Art Goebel could supply the answer. Art had started out as a barnstorming movie stunt pilot and had gone on from there to win the Pacific Dole Race to Honolulu. He retained important contacts in the film industry, so he might be able to help. In any case, I wanted him to fly the Sunbeam himself first. I knew he'd give me the unvarnished truth as to whether it was me, the airplane, or a combination of both. I didn't have long to wait.

The next day he took the Sunbeam up alone and put her through the wringer for about fifteen minutes. He landed, got out, and beckoned to me to follow him over to his car.

"Is there any way you can dump that heap?" he asked anxiously.

"Not now, there isn't. We've run through too much of the original backing already just keeping the Pigeon in the air. I'm seeing Richfield Oil tomorrow to see about pumping in some more cash to keep us going, but I doubt if they'd go for another airplane."

He fiddled with his helmet strap.

"I don't see how in hell you're going to handle that thing over those mountains at night when she's heavily loaded." He nodded in the direction of the surrounding peaks. "However, to ease your mind, the flight problems aren't yours. She's just a miserably unstable airplane, and I don't see how you've been able to do *any* aerial refueling with her, let alone get as far as you have. You and Paul deserve decorations for making flight contacts in two airplanes as totally unsynchronized as these. And as far as I can see, no amount of aerial photography is going to disclose anything else. Both of you must be operating on instinct. Incidentally, did you see me kick her out of a ground loop just now? How about the Sunbeam people? Would they consider widening that landing gear or setting it back some? It would help an awful lot."

I shook my head. I'd already gone that route and been told rather loftily that *good* pilots didn't request expensive, nonessential changes. Translation: They were well aware of the problem, but any more lip from me would result in stinging criticism of my flying ability. "Pilot error" was the catchall for anything that went wrong with airplanes in the twenties. (Tex Bohannon once said that while he was alone in the air on his way from Los Angeles to El Paso, a storm blew in over Omaha, and the papers credited the wind damage in Nebraska to his prop wash! An exaggeration to be sure, but it summed up the situation rather neatly.)

Art's words had warmed my heart and bolstered my sagging confidence to the point that I was ready to take off again that afternoon. But when we visited the hangar we found the Pigeon in her favorite position, squatting serenely on the concrete apron with her entrails spread out around her while Paul sought vainly to cannibalize parts from other aircraft. Art pulled out a notebook and started busily writing down names and addresses of people he thought could help Paul. Handing the slip of paper to him, he turned to me and said, "Look here, you've been so busy keeping your head down and your eye on the ball that you're looking a little peaked. There's a real good hoedown coming off this Saturday night at the Roosevelt Hotel. How about coming with me? Do you good to get out of all this fresh air!" I had no idea what a hoedown was, but if he'd said he knew a nice place where everybody walked the plank at midnight, I'd have jumped at the chance.

Because he had been the first aviator to conquer the Pacific, Art Goebel's fame rivaled Lindbergh's. On top of that, he was as tall and darkly handsome as any movie star. He was unmarried, a good bit older than I, and our friendship revolved solely around planes and flying. But no one else knew that, so I relished the chance to see the movie starlets eat their hearts out when I turned up on his arm.

That Saturday night I also discovered how kindhearted and thoughtful he was when the "hoedown" turned out to be a party he arranged in my honor, composed of mutual friends I'd been unable to see because of my flight commitments and living so far out of town at the Trouts'. I was delighted to see Bob and Viola Loutt (Bob was the Kendall Oil representative and an old friend from the Waco days), B.F.

and Tommy Mahoney (of Lindbergh-Ryan-Mahoney fame), the Fred Fosters (my good friends in Richfield Oil), and Roscoe Turner — minus his pet lion, but with a stunning blonde on his arm. Even Freeport was represented when my friends and neighbors the Leo Carrillos, the Charlie Macks (of Moran and Mack, The Two Black Crows, fame), the Charles Middletons, Tom Dugan, and Jimmy Conlin stopped by. Middleton, Dugan, and Conlin eventually became very important character actors. In those days, Middleton played Abe Lincoln in every Hollywood version of the great man's life, and Dugan and Conlin both were comedians who became important character actors. Leo Carrillo was descended from one of California's early wealthy families and was a star in his own right. He was quite an athlete and had taught me to swim the backstroke one day at the Casino Beach in Freeport, an occasion he referred to in a flattering little toast he proposed in my honor.

In the midst of all this fun and confusion, Art Goebel brought over a short Prussian-looking bemedaled stranger who Art said was Germany's top surviving war ace from World War I. The man seemed shy and out of place, but when the guests resumed their seats after Leo's toast, he stood up and called for them to drink to "the outstanding female pilot of our day — a mere child of eighteen, she has ..." and he proceeded to drone on in the most flowery and ridiculously untrue assessment of my flying activities that could be imagined. I wanted the earth to open up and swallow me, and things didn't improve when I heard him mutter *"Ach, Gott!* Such lies! Eighteen she says — twenty-eight is more like it! And they say this one *flies?* With one foot on the ground, maybe ..."

I looked around for Art and was amazed to find him doubled over with laughter, as were most of the other guests. Our "Ace" was none other than Vince Barnett, professional ribber par excellence. Seeing my agonized look, he apologized profusely and said he would do penance in any way I suggested. On such short notice the best I could come up with was to ask the group to raise their glasses to "the biggest phony in the kaiser's air force!" Such was the beginning of one of the strongest and most enduring of my friendships formed in Hollywood.

That party at the Roosevelt Hotel was the forerunner of many good Saturday night entertainments. It was the custom for all of the film colony greats to take the floor for some impromptu entertainment. Depending on the sobriety of the performers, the results were sometimes less than spectacular. But on other occasions we were pleasantly surprised when the entertainers displayed hitherto unsuspected talents.

Charles Bickford, one of the screen's meanest heavies, possessed a rich baritone voice and loved to sing romantic ballads that had the ladies swooning. Another surprise was Sue Carol, who later married Alan Ladd and, later still, became one of Hollywood's more astute agents. She danced and sang up a storm, things I never saw her do in a film. Clara Bow couldn't sing at all, but she managed to serenade Harry Richman, her love of the moment, with a noisy rendition of "I Wanna Be Loved by You Boop-Boop-a-Doop!" Jack Benny, then a handsome young man with prematurely gray hair, always did a timely monologue and played his violin. Janet Gaynor sang "Sunny Side Up," and Marion Davies turned up on several occasions to show everyone what a magnificent set of matched emeralds *really* looked like.

It was the closest thing to the Freeport LIGHTS Club one could imagine, and I felt thoroughly at home. I invited Bobbie Trout to accompany us several times, but her interests were not theatrical, and I found this was something we could not share.

Almost by accident I discovered that Viola

Loutt was the daughter of James Oliver Curwood, a famous novelist whose tales of adventure in the great outdoors rivaled those of Zane Grey and Jack London. Some of his books had been made into films, and through her father, Viola knew or was known to most of the upper echelon of the movie colony. Bobbie and I had had our pictures taken (through the efforts of George Ullman) with quite a few Hollywood celebrities, among them Bebe Daniels, Ben Lyon, King Vidor, and Anita Page, but I was no nearer being invited inside a studio than we'd been on my first day in town. Viola decided to rectify that and shortly afterward called to ask if I felt up to a trip out to Culver City, where they were filming something called *Anna Christie* with Greta Garbo and Marie Dressler.

I would have rowed a skiff to Tasmania to get a close look at Greta Garbo, but Viola warned me that she probably wouldn't appear that day, particularly if she learned there would be visitors. I still don't know whether it just happened or was carefully planned, but Garbo *did* appear, and she was most gracious. She was quite tall, had the most astonishingly thick real eyelashes I have ever seen, and did *not* have big feet, as was commonly rumored. Aviators were still so much of a novelty that she professed to be happy to meet me. I was incapable of a rejoinder, so Viola took over and did the honors, including my introduction to Miss Dressler. Marie Dressler had been a stage star of formidable dimensions and, according to Father, was one of the theater's most talented performers. But time had somehow passed her by, and she was now in her sixties, taking roles that producers wouldn't have dared offer her in the past. She warmed my heart by asking about "Tom," and we chatted until she was called to the set.

The scene was supposed to be taking place on an old barge at dockside. Actually it was set up on what looked like the sawed-off front end of a garbage scow. The "barge" rested in a large muddy puddle that two grips lying on their stomachs were agitating with large flat sticks to simulate the gentle lapping of moving water at pierside.

Garbo's voice didn't carry to where we were permitted to stand, but Dressler could be heard a block away. And this didn't change even when sulfurous torches were placed about the set to induce what would appear on the screen as a creeping fog. The fumes were so strong that wet cloths were passed around to be placed over mouths and noses to allow breathing. Dressler never missed a beat. She was star quality all the way. After the release of *Anna Christie* she became a bigger attraction than ever, regaining her starring status by playing everything from dignified dowagers to the unforgettable Tugboat Annie.

Still on cloud nine, I came back to the Trouts' from the studio to find several disturbing letters from home. Bobbie and I were being interviewed and photographed so constantly that Mother had made it a practice to enclose clippings from the New York papers. Her letters were always chatty and informative, but of late there was never any mention of Father's doings or his itinerary for the winter. Joe's letter was more explicit. It seemed that the United Booking Office was holding out on his contract (in those days, each theater required a separate one, even if only for one week's booking) until he would agree to have me join his act as an added attraction. Ullman's publicity campaign had done its work so well that United obviously planned to cash in on it.

Father was holding out, for which I was very grateful. I could just picture the grist my joining the act would make for Putnam's mill since he'd invented the fable that the stage was where I was headed all along. But Father understood perfectly that if he knuckled under to the Keith office demands, I would never be taken seriously in the aviation world again. However, the situation presented serious financial problems for Father. Because of the time lost on *Golden Dawn* the year before, he hadn't been able to put in a full season on the Orpheum Circuit. We were now well into November, and there were no bookings in sight. I couldn't recall this ever happening before.

I called home, and of course, both he and Mother assured me that I was not to worry, things would work out, they always had – etc., etc. Besides, they knew I had plenty of things to think about regarding the flight, and I was not to take my mind off that for one instant. They wanted me back in one piece – everything was just fine.

Out at Metropolitan, the Pigeon was laid up again, so Vince Barnett and I booked ourselves into some local air shows for what Vince called walking-around money. Each time I was featured on a program, I'd bring him along in a different bemedaled uniform and introduce him as a World War I ace from whatever country we happened to decide on. He was a master of dialects, so his commentary on my aerobatic exhibitions always broke up the show. I was very proud of the first $500 I was able to send home that I earned this way, even though I knew it was only a stopgap. If the Keith office were really about to starve Father out, he would need a steady income for as long as it took.

The contract Bobbie and I had with Ullman would produce substantial sums of money only if we set a record exceeding 100 hours, and the way the Pigeon was behaving there was no way to predict when that would be. To complicate matters more, the great Wall Street crash of 1929 had come and gone. The full impact of this financial debacle wasn't felt for some months by a large segment of the public (the noninvestors), but I quickly learned that the big companies like Richfield Oil, whose coffers had heretofore been open to me, were far more cautious with their "miscellaneous" funds. The established practice of borrowing cash against future performances was now scrutinized far more carefully.

I took my troubles to Art and Roscoe. Roscoe was all for asking Howard Hughes to take me on to do some of the long shots in aerial scenes they were currently reshooting. "Hell, kid," said Roscoe, "you can't do the dogfights or any of that stuff, but there's still plenty of work away from the main cameras that you could handle as well as any of the barnstormers that drift in and out." Art pointed out certain difficulties with the Motion Picture Pilots'

Association that would have to be surmounted, but he conceded that Hughes had enough clout to do just about anything he wanted to, so Roscoe set up an appointment.

Time has erased from my memory just why I went to see the eccentric young millionaire in my street clothes rather than flying gear. It was an unfortunate gaffe. Hughes proceeded to treat me like some aspiring starlet who was begging for a chance to appear in one of his productions. Well, that may be a bit of an exaggeration. He certainly didn't chase me around the room, but he was boorishly surly, barely raising his eyes from his desk. I got out of there in a hurry and kept a luncheon appointment with Art.

He was very annoyed at Hughes but he had a better idea if I would just be patient for a few days. It was well worth the wait. What came out of that was a reintroduction to Clarence Brown, the eminent MGM director and private pilot I'd already met once on the day we went to see *Anna Christie* being made. This meeting led to about five weeks' worth doubling in flying shots for Dorothy Mackaill and Anita Page for $750 a week. With false blond curls stitched into my helmet and my goggles over my face, I was totally unrecognizable. All I had to do was to fly and stunt within camera range, then land in front of the camera and occasionally wave. It was the easiest money I'd ever made, and I was particularly lucky never to run into a scheduling problem. I never had to fly every single day in either our endurance test runs or in this.

How Clarence Brown got past the Motion Picture Pilots' Association I never knew, and I didn't ask. But I was warned repeatedly that if anyone became aware of what we were doing, it was all over for me and would cause severe embarrassment to him. I couldn't risk telling even Bobbie, with the result that the secrecy surrounding my movements on the days I didn't check in at Metropolitan drove a wedge between us that was unfortunate.

On top of everything else, I was becoming increasingly uncomfortable at the additional strain I was imposing on Mrs. Trout as the flight dragged on with its interminable forced landings. From being an overnight guest I had become an added member of the household. Keenly aware of the extra cleaning, laundering, cooking, etc., that this burdened her with, I saw no way to lighten the load as long as I stayed in her home. We were much too busy and involved with the endurance flight alone, to say nothing of the extra chores I had taken on in Culver City.

Viola and Bob Loutt finally rented a tiny bungalow for me in their court in Santa Monica. I told Mrs. Trout that my old friends from the East had prevailed upon me to stay with them for a while as they had plenty of room and available domestic help (not true), and I considered the change a good one since I would no longer be imposing on her (all too true). A bond of affection had developed between us, and it wasn't an easy parting on either side.

Things out at Metropolitan were still not going the way we'd hoped. Our flights were lasting from ten to eighteen hours, but the Pigeon's average seemed to be about three refueling contacts before she conked out again. Good news from home raised my spirits, though, for the Keith office had finally decided that Father meant what he said and had booked him solidly through June 1930.

By late November Bobbie and I knew it was now or never. The rainy season would be starting shortly, and there was no way we could stay up for a week to fly through that. Somehow the very finality of its being our very last shot at a world's record — for this year at least — gave us all new hope. Even the Pigeon seemed to give her feathers a determined shake as Paul taxied her out for a final test run. For the first time, the Sunbeam took off with a full load of fuel as gracefully as an eagle leaving its aerie. There was no aerodynamic explanation for this. Weather and runway conditions were exactly the same. But that is sometimes the mood of airplanes, as any experienced pilot will testify. I was so delighted at her change of attitude that even if there *had* been a technical reason for it, I wouldn't have cared. I just wanted to take advantage of her good humor. When we passed the twenty-four-hour mark, our excitement grew, and when we made thirty-six, we were sure we were on our way. These high hopes were dashed when our contact at sundown of the second day was abruptly broken off as the Pigeon's prop arced to a stop. She sank so fast this time that Pete was unable to get the hose coiled back up in the fuselage, and Paul was forced to land in a tilled field with the hose still streaming out behind him.

We had only taken on a little more than half the gas needed to see us through the night, but we decided to try to stretch our supply and gamble on making it until dawn. We knew that Paul would get up to us then if there was any way in the world to get the Pigeon off the ground.

About four in the morning the motor gave a cautious cough, alerting me that it was getting dry. I switched onto the emergency supply, shoved the nose down, and headed back to Metropolitan. One of the now-familiar fogs was rolling in under us, and I got there just in time to see the red runway lights disappear into it. Blipping the engine to signal those below that we were in trouble, I made a pass over what I hoped was the administration building and roared back up into the soupy overcast. I had never actually used the emergency system. According to the engineers, I should have at least a twenty-minute supply, but in a spot as tight as this, I preferred to lean on Smith's Law and figure that as sure as next Tuesday, things were going to get worse before they got better.

The gleam of boundary lights through some wispy fog sections, plus a newly started-up re-volving beacon, was a welcome sight. And some inspired soul blinked the red runway lights at just the moment the engine gave a mournful gasp and died. So much for engineer-ing data. I figured out later that the emergency gas lasted for four minutes — not twenty, as advertised. But a light breeze had sprung up and momentarily dispersed the fog to where I could line myself up with the field's boundaries. I guided the Sunbeam in over the wires I couldn't see but knew were there. A quick side-slip to gain speed so she couldn't stall out under us, a sharp yank to pull the nose up slightly, and she dropped in from about ten feet off the ground. I was taking no chances on any of her ground-looping quirks at this late date, but in all fairness to the Sunbeam, she took the shock like a veteran. One good bounce, and she was down for good. We had been up exactly 42½ hours, and the first official refueling endurance record for women had been established.

Bobbie handled the reporters so skillfully that none was aware that it was all I could do to work up a weak smile for the photographers. Amid all the congratulations and backslapping, I did my best to hide my keen disappointment. I was tired and probably overreacting, but to me a record of 42½ hours, when we'd worked so hard trying to stay up for 168, represented a crushing defeat.

Bob, Viola, and Vince loyally stood by all during that last night. They had been at home celebrating our thirty-six-hour success when the radio blared out the news of Paul's forced landing. According to Bob, "We hightailed it out to Metropolitan to get the old Travel Air warmed up and ready. Our original idea was to lead you back to the field if you were running low on gas and the fog drifted in, but so many other guys had the same thought that we had to start pulling ships off the flight line to give you plenty of room to get down if you had to come in in the meantime. When we heard the motor cut that first time, we couldn't see you at all, and then everything happened so fast, we couldn't have done anything to help you anyway."

Even so, their very presence was heartwarming, and I was overwhelmed by their concern. I was also very grateful to the Almighty and offered up silent prayers of thanks during the drive back to Santa Monica. There were so many other things that could have gone wrong and didn't. If the gas hadn't lasted to get us back to Metropolitan . . . if the fog hadn't lifted at just the right moment so we could see the field . . . if we hadn't had enough flying speed left to clear the wires. . .if. . .if. . .if.

11

Getting High with the Navy

Vince Barnett tipped the bag-laden Twentieth Century porter and turned to face me in the small train compartment.

"I wish to God you'd level with me on what this is all about. Why are you sneaking out of town like a criminal, for chrissakes? What'd you do, knock over a bank or something when I wasn't looking?" He started ticking off on his fingers: "One, Viola's mad as a hatter. She got you a top picture agent you wouldn't even go to see. Two, Leo Morrison told me he's got a solid picture offer at fifteen hundred dollars a week with escalator clauses. And three, you and I are set up for at least half a dozen of those air show displays as soon as the rains let up. And this time, with another brand-new world's record under your belt, we'd be bound to clear at least a thousand dollars a shot.

"But no," he continued, in a voice heavy with sarcasm, *"you've* got to go home — and by train, yet. What in hell's wrong with T.A.T.? Think of all the publicity they'd drum up across the country. Pictures and interviews at every stop. A shot of you at the cockpit controls — I just can't figure you. Are you pulling a Garbo on us or what?"

"No, Vince," I said quietly. "You've all been wonderful to me out here — but it's getting too show-bizzy for me. If I made a movie, I'd never be taken seriously in the aviation field again."

"But you might be a big hit," he protested.

"Not when they found out I couldn't act — and you know I can't, Vince. Besides, even if I could, I guess the plain truth is that I don't want to. Flying is my life, and I want to get back to it."

"Well," he said, scratching his bald pate, "if there's no way to change your mind . . . but tell me one thing. Just where did you get the idea that you can't act? What about all those straight-man introductions you gave me at air shows and dinners? How about all those times you had to wing it with the ad libs when there were last-minute changes? You don't call that act-ing?"

"No, Vincenzo, I don't. I was simply having a wonderful time and got carried away with the spirit of the moment."

He turned away sadly. "Well, so be it. By the way, what have you got lined up back East?"

"Absolutely nothing. I don't even know if I can get a job back there, but I have to get back to my beginnings. Can you understand that?"

"No, but I haven't got a choice. And you still haven't enlightened me on why this retreat from the world has to take place on a train."

"You just said why a moment ago. It's the one way to avoid all pictures and interviews, until I get my own feelings sorted out."

He looked down at me and shrugged. "Women! Who can figure 'em? Promise me one thing, though. If things don't pan out back there, you will let me know, won't you?"

"Would I dare not to? And, Vince, I'll never be able to thank you for all the good times and fun we had."

Raising clenched fists over his head, he went into his best Red Baron accent. *"Gott in Himmel!* Such a craziness . . . we haff such goot times she must go avay. Maybe next time ve haff bad times undt she shtays, ja?"

We both broke up, but the "All ab-o-o-o-o-ard!"

cut short our good-byes to a single gruff hug and his murmured "God be with you, little one" in my ear.

I had plenty of time for reflection on that long trip home, and I needed every moment of it. Heading the list was the situation between Bobbie and me. The need for secrecy during the time I was earning money for the family at home had driven a wedge between us that I was unable to bridge or explain. I regretted it sorely — and always would.

My disappointed reaction to the refueling endurance record was anything but sporting, and that needed the wrinkles ironed out of it, too. Instead of being eternally grateful that the four of us had beaten the odds against us, I found myself resenting the fact that all the courage and skill of my three colleagues had not been enough to keep the venture from being so drastically curtailed.

I also regretted jeopardizing my friendship with Viola, and looked forward to making my peace with her. In many ways she was a younger version of Lady Mary, hammering away at me that no matter how well, how high, how fast, or how long I flew, there was no way I'd make it to the top until I amassed some *real* money or married it. "It's the only way for you," she warned. "You've gone too far to turn back now, and you simply must have more financial independence."

When I looked at it logically, the both were right. Still, acting logically had never been one of my strong points. What, for instance, was logical about flying under those bridges? Yet everything that had happened stemmed from it.

It has been said that each of us marches to the sound of a different drummer, and I was fairly sure that this was at the bottom of my difficulties with my well-meaning friends. I understood perfectly the kind of success they wanted for me – and I was grateful for their concern. But I was too inarticulate (and probably too young) to express adequately my feelings to them where flying was concerned. I wanted someday to make a contribution to this business I loved so much. In order to be able to do that, I had to keep my skills at top pitch at all times. I couldn't take time off to make movies, go on personal appearance tours, and participate in all the rest of the folderol that went with that kind of triumphant achievement. But the bottom line was that deep down, I knew I hadn't earned it yet. Maybe I never would, but until I did, I'd just have to follow my own Yellow Brick Road to my own Emerald City.

One of my deepest regrets of the past year

was not having been on Mitchel Field back in September to see Jimmy Doolittle make the world's first blind flight. With the plane's cockpit completely hooded over, he had taken off, circled, and landed without ever seeing anything but the guiding instruments on the panel in front of him. In 1929 pilots were far from convinced of the infallibility of instrument navigation, so this demonstration gave the aviation industry an enormous boost in confidence. I am still awed at the courage and skill that Doolittle demonstrated by making that flight. I am also surprised at the lack of attention it gets in today's history books, for that flight laid the foundation for today's airline travel, which, in turn, has affected the lives of everyone on this planet Earth.

There had been no time to go through my mail before leaving Los Angeles, so Viola had packed it all in a small suitcase, which I opened after I left Chicago. My correspondence with Lady Mary had been limited to short notes she was still dictating to a secretary. This last one was postmarked two weeks before in New York. She began her letter by saying brightly, "They are sending me up to Boston for a bit of face changing. Who knows? I may come out of this looking like one of your film stars! I'll have to stay there awhile — six weeks or more. Do you think you might be able to come up for a visit? We could have such a lovely chat ..." This was followed by a couple of pages of aviation gossip which I quickly put aside. Face changing? She'd never said anything was wrong with her face. I knew about the skull fractures and her eyes, but no one I knew had actually seen her with the bandages removed. I decided Boston would be my first port of call after spending Christmas with the family. The next letter was from G.M., twitting me gently about the RCA fiasco and extending his usual invitation to come down to New Castle, an invitation made even more attractive by a hint at some new and secret designs he was working on.

A pleasant surprise awaited me at Grand Central Station where I was greeted by my attorney. James E. Smith, Jr., was the son of a former New York City district attorney and the grandson of a Rhode Island senator. His family had a summer home in Freeport, and he was a great fan of Father's. He became Father's attorney and also represented me whenever necessary.

As I had moved more and more into the public eyes, Mayor Jimmy Walker's office frequently extended official invitations to functions appropriate to my flying interests. Greatly flattered, I had never accepted any of them, feeling out of place in the political arena. When Jim Smith found out about this, he convinced my parents that it was important to broaden my activities and offered his services as an escort. Since he knew everyone from the governor of the State of New York to the lowliest clerk in the Hall of Records (and all on a first-name basis), I subsequently often found myself in the midst of a most enjoyable social whirl.

On this particular day he had assembled a whole group of young people with whom we had attended these political soirées. After three and a half days of my own company on the train I was delighted to see them all. Jim had coordinated everything with my parents, so after a quick call home we all piled into the waiting cars and drove out to Long Island. He had arranged with a restaurateur-client in Old Westbury for a delicious meal which awaited our arrival. For the next few hours I went right back to being an eighteen-year-old girl who loved pretty clothes, parties, and beaux, not necessarily and not always in that order.

After a few days at home I found I had absolutely no desire to visit the field. The memories of Red had been too painful the last time I was there, and there was no reason to feel they'd be less so now. The weather was terrible right through Christmas, so I decided to make the trip to Boston to see Lady Mary by train. When I mentioned my plans to Jim, he decided to accompany me and complete some business up there he'd been putting off. In this way we could have dinner in "Beantown" and return the same night.

Lady Mary was in a small private hospital the name of which I have forgotten. I was shocked when I saw her. One of her eyes didn't focus, and it was obvious that her cheekbone had been shattered. She insisted she was in "absolutely superb" condition and couldn't wait to get at the controls again. The nurse avoided my questioning look and quickly left the room.

Lady Mary leaned toward me and in a stage whisper that could be heard down the hall, said, "Gloomy Gus! Keeps telling me not to get my hopes up. What does she expect me to do with them, shove them underground? Annoys me terrible because, you see, I'm *really* so much better, m'dear." Reverting to a normal tone, she asked pointedly, "And now, what about you? No millionaires yet, I see. But tell me all about your flight with Bobbie first. I want to know *everything*!"

I was about halfway through this recital when Gloomy Gus reappeared with a vial of nauseating-looking stuff she insisted her patient drink. Despite Lady Mary's vehement protests, she stood by until every drop had disappeared. Lady Mary pleaded, "There now, I've been ever so good, haven't I? Now please go away and don't come back for a while. We are so looking forward to a nice long visit..." but Gloomy Gus led her patient back to bed and motioned me to wait in the hallway.

When the nurse closed the door softly behind her, I was amazed to hear a slight but unmistakable snore. "I'm very sorry," said the nurse, "but Lady Heath is much weaker than she lets on. To make things worse, she sleeps very poorly. We have to give her the strongest drug she can stand to knock her out. There really isn't much chance for the plastic work to improve her appearance, you know. But we are even more concerned with her private worries and the effects they may have."

Convinced that I was a close enough friend to be told about those, she said that Sir James was not finding it convenient to concern himself with a seriously injured wife so many thousands of miles from home. In fact, if the nurse had her facts straight (and I was sure she had), there was every indication that a divorce was in the offing.

I wasn't too surprised, in light of Sir James's

original decision not to come to America in September at the time of her Cleveland crash. Bill Lancaster still flew up every week to see her, so I decided to get the rest of the story from him. The nurse said that the potion she'd given Lady Mary would keep her asleep for several hours; that meant I couldn't wait until she awoke. I wrote her a chatty letter, called Jim at his client's office, and left the hospital to meet him for dinner.

During the meal and all the way home we talked of nothing else. With all his political contacts, I was hopeful there was some way Jim could help her. He promised to do his best but pointed out a few of the difficulties. The lawsuits involving the Great Lakes Company were undoubtedly being worked out in Ohio. And whether Sir James actually sued for divorce in England or South Africa, it was going to take sizable sums of money to work up a defense or a series of delays. Jim was only ten years older than I, but sometimes I felt as though I had another father. At that moment I didn't want to hear why I couldn't help her, but where and how I could. We were still no nearer to a solution — other than my decision to talk to Bill Lancaster — by the time we arrived back in Freeport that night.

Before I managed to locate Bill, G.M. called

asking if I could come down to New Castle right away. I started to tell him about Bill and Lady Mary, but he indicated that this was a matter that had surfaced since his letter and was of an extremely urgent nature.

The Waco was up on blocks for the winter, but Sonny Harris lent me a Travel Air, and I left early the next day. G.M. was his usual gracious self, insisting that we lunch first and then get down to business. I liked the sound of that. Since my return the only offers of substance I'd received involved either more endurance flying or making a national tour to promote a new small open-cockpit biplane. I felt strongly that everything necessary to prove the performance of both pilots and airplanes under long-term varying conditions had been pretty well confirmed. Any more endurance flying would, I felt, be looked upon by the public as stunt flying, and I wanted no more of it. I did want to see a female team set up a refueling record that would compare favorably with the current male record, but I personally was not interested in going after it again.

The light plane tour was very attractive financially, but I really wanted to stay in the large monoplane category if I could. Over some delicious she-crab soup, I found out all over again why Guiseppe Mario Bellanca was always so far out in front of his competition.

As he put it, "I was looking into my crystal ball, and I saw the most marvelous thing. Instead of battling their way through storms and bad weather, all the airplanes were flying high above all these turbulences. Even birds," he said, his dark eyes twinkling, "don't try to fly at such times but seek shelter until the disturbances pass. Only man pits himself against nature's forces. Why do you suppose he does this?"

"Because," I said, breaking up some pilot crackers, "if he wants to set up a successful airline, he has to attract passengers away from the railroads, which aren't troubled by bad weather. He's selling them speed and service and making every attempt at dependability."

"Exactly," said G.M., laughing delightedly as if I'd just refuted Einstein's theory. "However, we have our solution to this, no?"

"We do?" I asked dubiously through a mouthful of crackers.

"Of course," he said patiently. "We fly over everything, up high where it is always calm."

I finished my soup and tried to think of a way to avoid making a comment. All the military penetrations of extreme altitude in both powered flight and free weather balloons had indicated an environment more hostile to man than the one found under the seas. He couldn't exist up in the very thin atmosphere without a constant and dependable oxygen flow. Temperatures of up to seventy-five degrees below zero had been recorded on the balloons, and our current cabin heaters weren't very effective at even 10,000 feet. So how could passengers possibly be carried at what I assumed he meant would be 35,000 feet or more?

Two things crossed my mind. He'd almost never been wrong, and even if he was this time, I was a bit edgy about crossing swords with the boss. Luckily the waiter chose this particular moment to remove the soup plates and serve our salads.

"It is certainly a fascinating concept," I said cautiously.

"It will be much more than that," he said enthusiastically. "We must have different engines, cabins will have to be insulated, reinforced, and – yes – pressurized. Even new metals will have to be discovered or invented before it becomes a practical reality. But first things first. We have to draw attention to its commercial possibilities, yes?" I was beginning to see that this was where I came in. "How would you like to do some high-altitude work?" he concluded.

"You mean *really* high, like Apollo and Zeus

Soucek, the Navy altitude aces — past, say, thirty-five thousand feet in some of that experimental navy aircraft?"

"No. I'm talking about commercial aircraft. Nothing that is flying passengers today can operate comfortably at over fifteen thousand feet, and few of them can get anywhere near that, even flying empty. But our new Sky-Rocket should hit thirty thousand feet easily. How about that for a start?"

Some start! The last time I looked, the highest any woman had ever gone was about 22,000 feet. I speared a shrimp and chewed on it thoughtfully. I wanted to do this one so badly I could taste it, but I didn't want to appear too anxious.

"You will train for it with George, of course," he went on, "provided you like the rest of the idea. This time I can't provide you with the airplane free of charge. My own backers are chary of the whole thing, and I have not been successful in getting them to see the commercial possibilities of this. But since you are now so well known, I hoped that one of the oil companies would pay the incidental expenses and fuel costs. I'm sure I can get Pratt and Whitney to donate a maintenance engineer's services. It is as much to its advantage as it is to ours that this be successful. You are still only eighteen, and you are much smaller than the average male pilot. If you are successful – and I see no reason that you should not be – I'm sure this will sway my own people."

We both knew was going to be a lot riding on this one. The mere mention of new metals, pressurized cabins, sophisticated power plants, etc. meant that his own backers fully realized the tremendous sums that would ultimately be necessary. It followed, therefore, that he was banking on me for nothing less than the full glare of New York newspaper and national newsreel publicity to highlight an unqualified success. That could be the only possible reason he hadn't assigned this important task to his number one test pilot, George Haldeman.

Fortunately George was working in the plant that day, so I had ample opportunity for an exploratory conversation. I was delighted to learn that George's recommendation had already been sought by G.M. and that it was what had locked in today's proposition. As with the endurance flight, George would give me the advanced training. Arrangements would be made by G.M. for the use of the Navy's high-altitude pressure chamber at Anacostia in order to satisfy all U.S. Navy requirements as to my physical condition.

I had never flown the Skyrocket, which was

larger and heavier than the Pacemaker and was powered with a 425 h.p. Wasp engine. I had only seen drawings of it and was unaware that it was ready for flight. Still, I knew that any airplane of his would climb like an eagle, so that was the least of my worries. I just didn't want to let anybody down.

I flew Sonny's Travel Air (which I later found he'd "borrowed" from his current employer) home, dodging the thickening "flurries" that turned into a full-scale snowstorm north of Trenton. Gritting my teeth against the bone-chilling cold, I heartily wished I had in the front seat that obnoxious weatherman who'd called this sleety mess a snow flurry when I'd called him from the New Castle airport. He was one of those superior types who resented my asking for prevailing winds and peripheral conditions. By the time I crossed New York's Lower Bay I was weaving between ship masts as I searched vainly for the Coney Island beach. I didn't dare go any higher, even if I'd been able to see, for I was already getting a slight build-up of ice on my wings' leading edges. I got down to about twenty feet over the waves and stayed there until the familiar buildings of Long Beach streaked by on my left. I followed the local roads back to Garden City and drew my first easy breath when the weather-beaten hangars of Roosevelt Field came into view.

I lined up my approach and started down. Suddenly I heard a familiar voice yell, *"Pull up — to your left — dammit!"* and to this day I swear I did not touch the throttle or controls as the faithful Travel Air zoomed into a climbing turn, missing by a hair's breadth the silver and yellow winged craft that was sideslipping directly into me. With shaking hands I completed my circle back up into the thickening snow and tried my approach once more. I got down safely this time and taxied back to the hangar, where Sonny was in a raging mood.

"You all right?" he barked harshly. "That SOB Roger Kahn almost shaved off your top wing out there. C'mon, we're going down to see that so-and-so and find out what in hell he thought he was doing with that new Corsair of his."

I was just as angry as Sonny, so we drove down the line to see Roger Wolff Kahn, son of Otto Kahn, the famous international banker. Young Roger was no Howard Hughes, but he had his own hangar filled with the very latest and best aircraft on the market. As far as anyone knew, he rarely flew cross-country in any of these aeronautical Rolls-Royces, but he was becoming a pilot of admirable skill since he flew practically every day. He paid his crew good wages, and they were very loyal to him. Today's performance was completely out of character, but the loss of the Travel Air could have cost Sonny his job and his license, to say nothing of taking my own life.

The lights burned brightly within the hangar, and we could hear voices coming from its little office. Sonny and I presented a strange appearance. Face still blue with cold, I was in my fur-lined flight suit and boots. Sonny's grease-spotted coveralls and heavy work boots were topped off with an old sheepskin-lined helmet that flapped over his ears. Kahn was a small man, not much taller than I, but Sonny's huge frame loomed over him, a towering fury.

"Who in hell do you think you are, cutting off a pilot like that?" he roared. Kahn was taken aback but assured Sonny there must be an explanation and, turning to one of his mechanics, said, "Send Lieutenant Carlson in here!"

Lieutenant Carlson was a neat and natty flyboy lent by the navy to check out Kahn on the brand-new Vought Corsair he'd just purchased. Under questioning, it developed that the lieutenant had decided on his own to take the Corsair up for a hop. "I'd been told to try her out under all weather conditions, sir." He addressed himself to Kahn, ignoring Sonny and me completely. His attitude was one of aggrieved arrogance — as if it were inconceivable that anyone should question his actions.

"Were you also told to watch out for other aircraft while in the air?" asked Sonny sarcastically.

"Who else would be in the air on a day like this?" Carlson shot back. "Unless, of course, it was a student who didn't know any better!"

Roughly pulling off my helmet so that my light hair cascaded about my shoulders, Sonny barked "Take a good look. Does *she* look like a student?" Sudden recognition glazed the young man's eyes, and he looked over at Kahn for support — which he didn't get.

"Let me tell you something, Mr. Smart Guy. I was watching from my hangar door, and if she hadn't dipped a wing and climbed up into that scud, we'd be scraping you off the snow with a shovel right now. This girl has been flying off this field since she was seven, and you reaped the benefit of every one of those years this afternoon. I've got a good mind to lodge a complaint against you with the Department of Commerce!"

Roger Kahn hadn't witnessed the incident, but he couldn't have been more solicitous in expressing his regrets. Sonny was somewhat mollified, but on the way back to the hangar he said he thought it might teach that young "smart ass" a lesson if he were called on the carpet by the department. I knew Sonny was probably right, but I was so grateful even to be alive that I just wanted to get home and thaw out. I also wanted to get off by myself to have a private talk with Red — wherever he was, because I certainly knew where he'd been an hour earlier.

My family wasn't crazy about this high-altitude opportunity when I talked it over with them, but they did understand my regarding it as the breakthrough I was looking for to get completely out of the "stunting" category. Joe was particularly sympathetic to this and was most helpful in swaying Mother and Father.

I decided to try the financial waters before getting back to G.M. I didn't want to see his disappointment if I wasn't successful, but I needn't have worried. The combination of G. M. Bellanca's reputation and the opportunity of setting a world's record for women in an area that had been entirely masculine up to that time was appealing. Richfield Oil and Champion Spark Plugs quickly picked up the initial tab for the testing period — about $5,000 — with promised sizable bonuses for me in the event of a record.

When I called New Castle with this good

news, I reminded G.M. of his promise to contact the navy and Admiral Moffett pointing out that we could move now as quickly as they could be ready for us. This last was extremely important, for when you got right down to it, all I really knew about this kind of high-altitude flying would fill three lines in a flight log. I knew George would be invaluable in getting the best possible performance out of the Skyrocket, but there are tricks in all trades, and I never forgot the difference the High Hats' training made in my endurance contact flying. I'd have probably gotten by on my own, but by the time they finished showing me the fine points of precision flying I was home free. I wanted that feeling again, and I was sure that the navy pilots could give it to me if they would.

I was packing for my stay in New Castle when Mother said Casey Jones was on the phone. I couldn't have been more surprised if she had said the Pope was calling from the Vatican, but I was too curious not to run for the phone. He was very businesslike, seeming quite disappointed that I couldn't drop everything and be in his office at the field at eleven o'clock (it was then ten thirty). I told him I was planning to leave town that afternoon, but if it was that important to him, I could meet him for lunch at one o'clock in a nearby restaurant on Old Country Road, alone. He wasn't at all pleased with that last condition, and I wasn't sure why I made it, but the old Irish ESP had taken over, and Casey's silence told me I'd struck a nerve.

George was in New York, and I was going to fly back down to Delaware with him, but I decided I didn't want Casey to know where I was going. He'd find out about it soon enough — in the newspapers. Knowing the element of surprise to the press would be lost if Jones got wind of our plans, George agreed to pick me up from the upper part of Roosevelt Field, where our takeoff couldn't be seen from the lower field hangar line.

Casey was old enough to be my father, so I was amused to find our conversation taking on the tone of a couple of contemporary horse traders. All he wanted was a letter from me absolving the Cessna Company from responsibility in the Devereaux crack-up! It was no secret that a million-dollar lawsuit by the Devereaux family was brewing. The value of such a letter from me, having flown and raced the Cessna — and survived — prior to Red's accident, would, of course, be very valuable. But Casey ran true to form. If I would agree to give him the letter, he would permit me to use one of the Curtiss

Robins any time I needed transportation!

Even now, fifty years later, I am amazed that he would even suggest my endorsing the airplane that was responsible for Red's death. My first instinct was to tell him off once and for all. But I knew it would be a waste of breath on anyone this insensitive.

I pointed out (with the greatest of pleasure) the circumstances of my earlier, unacknowledged letter to the Cessna Airplane Company. Also, the lack of proof that he *or* Cessna had ever made good on their promises to Red to look into the matter of the ailerons would certainly deter me from making any moves in their direction, even if I didn't feel as strongly about the whole thing as I did. As for the Robin, I was now in a position of getting cash endorsements for any planes I flew, so I'd have to pass on that one in any case.

"Oh, well, if it's *money* that's important to you, I'll pursue it." His patronizing air was almost more than I could endure, so I named a figure far in excess of what I'd been offered by anyone else and drew our meeting to a close. Obviously Curtiss wanted me to fly the Robin, but in Casey's convoluted thinking he'd put the two widely separated matters together. In any case I was sure that it was the last I'd be seeing of Casey Jones. I couldn't have been more wrong.

George and I left for New Castle at two thirty that afternoon, and by four thirty my training for high-altitude climbing techniques was in full swing. There were last-minute changes being made on the Skyrocket, so we worked in the Pacemaker. We wore oxygen masks (which I hated), and most of our flying took place at over 20,000 feet. It was important to get the maximum performance in the least possible time in order to cut down on the weight of the gasoline to be carried. With gas weighing six pounds to the gallon and my own weight — or lack of it — we were planning to offset the advantages of other aircraft manufacturers who would have to use men weighing much more than my 112 pounds.

George was patient and thorough, stressing the importance of keeping the motor in steady synchronization and functioning in the extreme cold. This was the only way to keep the ship in a steady climb. It was also important to be alert to any "mushing" tendencies that would precede a stall. There was plenty of altitude to recover in case it should occur, but precious fuel would be burning off while I was regaining climb attitude.

In a few days word came that the navy was ready for me at Anacostia, so I took off for the

Washington Naval Air Station. On hand to greet me were Miles Browning and his close friend Lieutenant Commander Ralph Ofstie, former Schneider Cup racing contender. The Schneider Cup was the trophy awarded the winner of this first international competition in speed racing. Ralph had a delightful sense of humor but occasionally got carried away and insisted to all that I was really a boy flying in "drag," wearing a blond wig. Privately he was another of the big-brother types who gave me valuable flying and racing tips, including the unvarnished truth about certain racing pilots whose sense of "honor" during competition was either warped or completely nonexistent. I'd won a lot of race money and avoided being chewed up by pursuing propellers by listening to him, so I was more than willing to play along with his teasing.

In honor of my appointment with Admiral Moffett, I had taken pains to dress carefully in spotless whipcords with highly polished field boots. My knit jersey blouse, scarf, and leather jacket were all in coordinating shades of powder blue that made my sun-streaked hair appear blonder than usual.

The effect on Miles and Ralph was all I could have wished. They made a mock search of the Bellanca's cabin insisting that no *Follies* girl like this could have flown in by herself to keep an appointment with the admiral. There must be some mistake. But just so I wouldn't be too disappointed, they'd use their influence on his aide and try to get him to give me a few moments of his time. At that moment, when they had me half convinced, the admiral's car drove up, and I grandly waved them off – but not before asking them to wait for me.

Ushered into Admiral Moffett's presence, I found myself just as much in awe of him as I'd been at our first meeting the year before. That had come about after a forced landing at Hampton Roads in one of Lady Mary's Cirrus Avians. Lady Mary had already left for Langley Field, but the engine problem that forced me down could be fixed only by new parts sent in from Canada.

The admiral had heard about my flying under the bridges and sent word through Miles, the base commander, that he would like to see me fly one of their training ships. Nervous and flattered, I got into this strange (to me) airplane and put on an aerobatic demonstration out over the waters of Hampton Bay. I noticed that it was pretty sluggish coming out of a precision spin, but the loops and rolls went well for that type of craft. Evidently that much of a show hadn't been expected, for I was told later that that particular trainer was strictly banned for aerobatics, particularly on spinning. There had been fatalities, when two of them simply hadn't come out. I probably got away with my spinning it because I didn't have any weight in the front seat.

When I landed, the admiral sent for me and, with a group of officers around him, took off his gold navy aviator's wings and solemnly pinned them on me. Then, as a further reward, he asked me if there was any other airplane on the base I'd like to fly. Talk about letting a child loose in a candy store. I couldn't think of any I *didn't* want a crack at. (Remember, this was taking place at a time when women weren't even allowed to be flown as passengers in military planes.) I finally settled on a brand-new beauty — a silver and yellow Vought Corsair, powered with a 425 h.p. Wasp engine, heavier than anything I'd ever sat behind.

I really thought that at the last minute I'd wind up in the front seat as a passenger, but the admiral was as good as his word. The Wasp was warmed up, Miles lifted me up into the rear seat, and off I went! If they hadn't been careful to limit me to a half tank of gas, I'd probably still be flying around Hampton Roads. It was a fantastic experience and a memory I shall always cherish.

Admiral Moffett brandished his patriotism like a bright sword. He believed the United States to be the greatest nation on the face of the earth, and he wanted every world aviation record in existence to reinforce his nation's worth in the eyes of the world. I knew he'd been severely criticized for letting pilots like Al Williams and Ralph Ofstie arrange their leaves and duties to coincide with international aerial competitions, but my civilian status would obviously require different handling. I thanked him profusely for the help extended by the High Hat Squadron from San Diego, but he modestly brushed my comments aside.

I made the mistake of thinking this a good time to bring up the subject of the treatment of women's records at the hands of the Fédération Aéronautique Internationale, but he sternly put me off.

"Miss Smith, it is always best to take things one step at a time. Let us start today with your assault on the world's altitude record. If you succeed in getting over thirty thousand feet in a nonmilitary aircraft, you will have won your own case. Right now, we can place the facilities of the Anacostia station with its pressure chamber and high-altitude flight equipment at your disposal. I have agreed to send a daily report to Mr. Bellanca on your progress here.

My aide has prepared your schedule. If you have any questions regarding it, please get back to me through him. Oh, just one more thing. I must ask you to be discreet where the press is concerned. There are important men in both the House and the Senate who do not approve of my expenditures of navy funds on skilled personnel in these record attempts. I trust you understand and will conduct yourself accordingly."

G.M. hadn't said anything about secrecy where the navy was concerned, and I couldn't help wondering if he was aware of what I'd just been told. Lunch with Miles brought things a bit more into focus. A certain New England senator was giving the admiral a very hard time on some naval appropriations. Moffett suspected that the senator was trying to force the admiral into giving Amelia Earhart the same kind of cooperation he had extended to me, because this senator was known to be a close personal friend of G. P. Putnam's. However, the admiral had seen Amelia fly at Cleveland and been singularly unimpressed. So far he had diplomatically sidestepped the Earhart issue and saved his appropriations bill.

It was the first I knew that Earhart was even interested in doing any high-altitude work. Actually she wasn't. Putnam was — the minute he'd heard through his grapevine about me. It was the same tired old story. If Amelia couldn't participate, he wanted to make sure nobody else could either. Only this time the buck didn't stop on an advertising executive's desk — it landed on that of one of the United States Navy's most respected admirals.

"Whew! And I thought I had troubles! But that part about the senator and Amelia seems a bit far-fetched, Miles. Are you really sure about that?"

"Well, all I can say is that I met the senator, G.P., and Amelia at a Washington reception about a month ago, and they all were on a first-name basis. And part of the conversation had to do with her obtaining a transport license — which, as you know, she hasn't been able to do up to now."

It certainly sounded as though G.P. were operating at full throttle. The transport ticket was held by only a few women pilots like Louise Thaden and Phoebe Omlie. It was the highest pilot rating obtainable. Having one entitled you to fly anything from Aeroncas to Ford Tri-Motors for hire and personal business use. According to Miles, G.P. was suavely insinuating to the senator that the government should be only too glad to issue one to "A.E." without her being put to the annoyance of the written, physical, and flight tests. But the senator shied away from that one, obviously afraid of public censure should the truth of the matter ever become aired.

I was definitely frustrated at not being able to acknowledge publicly the U.S. Navy's assistance. I had planned on it to strengthen my position in the upcoming altercation with the FAI. But I still had much to be grateful for. The admiral had not backed down on his promise to G.M. and – provided I passed all my tests successfully – the admiral's reports would undoubtedly calm down those of G.M.'s backers who still resented putting their equipment in the hands of a pilot who had been uninsurable up to now because of her youth.

I found the high-altitude pressure chamber very interesting. You could be pressurized to any altitude. You were given a pad and pencil with a series of problems to solve, and the oxygen was gradually reduced to monitor your performance under varying conditions. The first time I entered it I was convinced that I knew exactly where I was and what I was doing, but when it was over, I was embarrassed to learn that I'd failed the simplest tests and even managed to misspell my own name. But my testers said I scored well since my system shouldn't have been operating at all by the time the pressure had been reduced that much! The physical tests went well, too, except for the discovery that I had extremely low blood pressure. In the years to come this was found to be a distinct plus as far as high-altitude work was concerned, but at the time this was not known, so the doctors tried everything to raise it, even encouraging me to smoke under a doctor's supervision. Nothing worked, however, and since I was disgustingly healthy in every other area, I was passed with "flying" colors.

The flight tests went well, except for the oxygen mask. The one I'd been using at New Castle had to be operated by pressured exhaling — something that completely exhausted me. I also suffered severe claustrophobia from it because it covered my nose and interfered with my vision. I asked the navy engineers for a light metal container that could be worn around the neck, using a narrow strap as a holder, with a small tube that could be taped to my mouth. They immediately got to work on it, and thirty years later — when I first flew in a T-33 jet trainer in 1960 — the darn contraption was still being used for the ejection oxygen system. So much for technological progress . . .

Before leaving Anacostia, I met the current high-altitude world's record holder, Lieutenant Apollo Soucek. Another old friend, Lieutenant

Al Williams, was also there, and we had a fine visit. The best thing about that particular group of officers was that there were no male chauvinists in the crowd. They just wanted to see the United States hang up another record, and if I was the one elected to do the job, that was good enough for them. I could have been three feet tall, with two heads or green hair, for all they cared. I flew back to Delaware in a state of elation. It isn't every day a girl is promised the assistance of the navy air arm.

At New Castle there was a change in plans. The scheduled launching of the Skyrocket was postponed when she had to go back to the drawing board for design changes. G.M. had already decided to go ahead with the Pacemaker, which was fine with me, even though I had looked forward to flying the Wasp. On the other hand, the Wasp was a good bit heavier than the Wright J-6 in the Pacemaker, so it probably all balanced out.

By now we were into March, the best month of the year for what we had in mind. Her brisk winds would be a valuable tool for a high-altitude climb, provided we could get an accurate weather forecast. It was important to carry as little fuel as possible, and George teased me about measuring it out in a teaspoon. The effects of the freezing level of oil on the controls, engine metals, and panel instruments were still largely unknown factors. As a precaution, every drop of oil was wiped from the wires leading to the ailerons and tail section. The inside of the cabin was stripped bare, right down to the fabric covering. It made for a chilly, rough flight back to New York, and now my graceful bird, instead of reaching for the ceiling like an eagle, flew like a Mack truck with the bends.

The National Aeronautic Association timer had already arrived by the time I landed at Roosevelt Field. He had brought along the official barograph, so everything was in readiness for the next day's attempt. A call to Doc Kimball at New York's weather station down at Battery Park confirmed that flying conditions in twenty-four hours would be made to order, provided the northwest winds advancing on us from Pennsylvania stayed on course.

March 10 dawned cold, clear, and gusty. Arriving at the Field, I took on enough gas for forty-five minutes of flight, while the barograph was being wired into place.

The navy warnings during my training about my oxygen flow caused me to check out the intake valve on my mask once more. I could only gamble that it would function as advertised. In 1930 there were no options. If it quit on me above 20,000 feet, I'd never know it. Instant sleep would take care of that.

The big monoplane bounded into the air joyously, her climb greatly assisted by the high winds. I kept her on an upward spiral that was fairly tight, to avoid being blown backward over the Atlantic. The engine purred like a contented cat until I reached 27,000 feet on the altimeter, at which point she gave a bit of a clatter. Miles had warned me about this, saying it usually preceded some icing up in the carburetor. I blipped it a few times to clear out any frozen particles and kept going until the altimeter showed 30,000 feet. Knowing that my true altitude was probably a few thousand feet less, I tried to inch her farther upward, but the sputtering motor gave warning that ice was probably forming in the gas lines. The thermometer outside the window read forty degrees below zero.

Reluctantly I pushed the stick forward. By now the gas was practically gone. I conserved it as best I could, but just as I lined up over the Mineola farmlands for my final approach, the engine coughed once and died for good. I had plenty of forward speed, so my dead-stick landing didn't present much of a problem — except the ground crew had to send the gas truck after me to help me out of the cockpit. It was just as well. Bundled up in my furlined suit, helmet, boots, gloves, and oxygen gear as I was, there was no way I could make a graceful exit from the airplane for the newsreel cameras.

I had learned to my sorrow that male pilots could look like deep-sea divers or humanoids with braided faces from Mars — and reporters loved it; the eerier, the better. But let one of us women have a hair out of place or a smudge on her nose, and it would be triumphantly hailed coast to coast as a reason for *"Women belonging in the home!"*

G.M. was wreathed in smiles as I arrived back on the truck, and Mrs. Bellanca presented me with another bouquet of red roses. It was a happy time for all. And this victory was especially sweet for me after the letdowns of the past six months.

At home it was the endurance record all over again, with flowers filling the rooms, phones ringing, and wires of congratulations streaming in. Among the first to arrive was Amelia's, followed by Bobbie's and Vince's. His was typical: "Gott in Himmel. For this you had to go back to New York? Out here the weather's better, and you could have gone higher — THE BARON HAS SPOKEN!" Mayor Walker's was a friendly "Congratulations, I always said you'd do fine if you stayed high up over those bridges."

A few days later I took the altitude ship back down to New Castle for a complete strip-down to learn what we could from the condition of parts, wires, and instruments. At the end of the day George Haldeman called me in the hangar that there was a long-distance call for me on the office phone. It was the National Broadcasting Company wanting to make sure of my appearance on the program sponsored by Champion Spark Plug Company at eight o'clock that night. My explanation that this was news to me and that I had no plans to fly back to New York for a week or more produced such sputtering on the other end of the line that I thought the connection was bad.

But it wasn't the connection. It seemed that somewhere in the fine print of the agreement I'd signed with Champion was a clause guaranteeing my appearance on its national radio program in the event that my flight resulted in a world's record. I hated to admit that when I signed that contract, I'd been so anxious to get my hands on the money to move the flight along that I hadn't read all of it as carefully as I should, so I assured them that if they would have someone meet me in a cab at the North Beach Airport at seven fifteen that evening, I'd make the broadcast in good time.

I raced out to the flight line and told George what had happened, and he immediately gassed up one of the monoplanes. In a matter of minutes I was airborne. Luckily the weather was perfect, and it was a smooth-as-silk flight all the way. The New York skyline was breathtaking, its buildings shimmering in the gold of a setting sun. North Beach (now La Guardia Airport) was then just a two-track cinder path at the water's edge. The narrow airstrip was deserted except for the presence of a yellow cab waiting at the west end. I landed quickly, and as I shut off the engine, a tall young man jumped out of the cab and introduced himself as Timothy Sullivan, representative of the broadcasting company.

As we entered the cab, he told the driver, "There's a ten-buck tip in it for you if you can get us up to Seven-ten Fifth Avenue in twenty minutes. Don't worry about traffic tickets. NBC will handle everything."

The cabby, a typical New Yorker, shrugged and said, "Oh, sure — and then the governor's going to give me a medal, right?" Nevertheless, he whizzed across the Queensboro Bridge, threaded his way unerringly through the evening theater traffic and deposited us in front of the NBC building with a laconic "Here y'are sport!"

He'd made the trip in such incredible time that Mr. Sullivan upped his tip another $5. Stuffing the bills in his jacket, the driver slammed the car door and called out, "Lots of luck tonight, Elinor. I'll be listening here in the cab – and listen, call me when you want to go back out to North Beach. It's on the house!"

He must have been listening to our conversation on the way to the studio, for I had quickly discovered that Mr. Sullivan knew little or nothing about the "Champion Sparkers" program on which I was to appear.

His job had been to deliver me on time to the show's producer, and when that mission was accomplished, he promptly disappeared. Phillips Carlin, the program's announcer, took one look at me in my soiled flight gear and went into immediate shock. "No, *no!* Not another one! I simply won't go through that again! Has this one had any experience at all? How on earth are we going to teach and rehearse her in time? It's seven forty-five and we go on the air in fifteen *minutes!*" His voice rose to a wail as his studio crew all stared at him blankly. Seeing my bewilderment at this panicky scene, one of these

353

kind souls filled me in on the cause of his distress. The stellar attraction for last week's show had been Martha Norelius, the Olympic swimmer, who had fainted dead away when confronted with the microphone.

I approached Mr. Carlin and suggested that instead of working from a prepared script, he tell me exactly what they wanted me to say and how long I'd have to say it in. His expression couldn't have conveyed more shock and horror than if I had just told him I'd stabbed his mother. But with the hands of the clock inexorably advancing toward eight o'clock, he really had little choice. Despairingly he thrust a script into my hands, saying that it would have to be read at even speed during an eight-minute period. One glance at it told me that its author knew no more about airplanes than I did about Olympic swimming – but I didn't tell Carlin that.

When the engineer signaled that we were on the air, Carlin rolled his eyes heavenward and stepped up to the mike as though facing a guillotine. But when he opened his mouth, his advertising pitch was delivered in the perfectly modulated tones that had made him one of radio's highest-paid announcers. When my turn came, I held my script in front of me and didn't read a word of it. My experiences in

California with Vince made this kind of ad-libbing a breeze. After all, what was easier than talking about the subject I loved most? At the engineer's signal I concluded my talk by thanking the audience for their attention and suggesting that they write me at the studio if there was anything else they would like to learn about high-altitude flying.

The whole thing was so easy I found it hard to believe that there was such an enormous audience out there. If I got three letters, I figured I'd be doing well. I fully expected another scene from Mr. Carlin once we were off the air and he realized I had junked his script. But since I hadn't fainted, fallen on the floor, or embarrassed him in front of a national radio audience, he apologized most profusely, and we parted on a friendly note.

When I left New Castle, I had asked George to phone home and ask Joe to drive into New York and pick me up. Dressed as I was, I didn't relish taking a train home to Freeport, and since I was leaving Delaware in such a hurry, I wasn't prepared to check into a hotel either.

Not to worry. As I left the studio, I saw Joe waiting in the hall. Right behind him was an NBC page who said he'd been instructed to escort me to Mr. Aylesworth's office. When I arrived at these impressive quarters, I found him flanked by two NBC vice-presidents and a male secretary. I couldn't help wondering what little surprise Aylesworth had in store for me this time. Conducting the introductions as though we'd never laid eyes on each other before, he got right down to business.

"Miss Smith, we find it difficult to believe you have never had previous radio experience. We all were very impressed with your performance tonight, which Phillips Carlin tells me was completely extemporaneous. We were wondering if you would care to come to work for NBC on a steady basis."

"Seems to me you asked me that once before," I said slowly.

Aylesworth had the grace to blush. "Hmmm, yes, well, that was most unfortunate. You can be sure *that* won't happen again. We have an option here for you to sign if you'd care to look it over?"

"Mr. Aylesworth, it's been a long day, and I'm really very tired. Maybe we'd better wait and see if there is any mail response to tonight's broadcast."

"Oh, that won't be necessary. We can predict a heavy mail flow right now. Aviation — and you — are of top interest at the moment.

Besides, we are not thinking only in terms of studio broadcasting but of shows to be done on our remote-control system directly from airports. Does that part of it appeal to you?"

"Yes, that part really does. But I think you'd best send those papers to my attorney, James E. Smith, Junior. His office is on Chambers Street. I must return the ship that I flew up tonight to the factory tomorrow. I'll be back in New York next week. Could we discuss it then?"

"By all means. In a week, then," he said, extending his hand.

Joe eased the car out from the curb onto a deserted Fifth Avenue. "So little sister wants no part of show biz, eh?" he chuckled.

"Listen, you know they'll have probably changed their minds completely by next week. You didn't expect me to stamp my feet and turn them down flat, did you?"

"No, but I didn't expect you to turn into a talent agent at the drop of a hat either."

"What do you mean, agent?"

"Well, if you're as good as they seem to think you are, maybe CBS will get into the act."

"You know, I'd love to take credit for thinking that far ahead, but I was so tired I just wanted to get out of there without signing anything I'd regret later on. CBS ... say, wouldn't *that* be something?"

12

Microphones and Maneuvers

By the end of the following week Joe was walking around taking bows and saying what a shame it was that nobody had ever asked his professional opinion before because CBS *had* responded to the announcement of my meeting with Aylesworth in the trade papers like a fire truck to a four-alarm blaze. It would be gratifying to report that this came about because my radio performances had been so outstandingly brilliant that the eyes of the broadcasting world were focused upon me, but the truth of the matter was something else altogether. CBS and Ted Husing (its star sports reporter) had been at odds for some time over his current salary. One of the vice-presidents there was convinced that Husing's salary was too high and that Ted should take on all the upcoming aviation events as well.

Husing didn't know too much about flying at the time, but he could "wing" anything until he caught up with the facts. As is usually the case in such altercations, said VP saw no reason why Ted's talents shouldn't be kept at the same fiscal level, despite the added pressures of learning enough about a whole new industry to broadcast it intelligently. It was unfair to expect it of him, and Ted was balking — and then I turned up on the "Champion Sparkers."

CBS's offer for my services was substantial enough to give Jim Smith plenty of clout in dealing with NBC, so by the time I got back to New York my finances had taken a decided upswing. In fact, they had never swung so high in my flying life. As a sports fan of Husing's from way back, I wasn't too happy to find myself in the middle of all this, but I knew that if I hadn't happened along, CBS would have found someone else to hang over his head. I could only hope that one day we could meet and be friends in spite of all the corporate shenanigans.

NBC had already lined up a program for me to be broadcast three times weekly under the sponsorship of the Daggett & Ramsdell cosmetics company by way of the McCann Erickson advertising agency. But this was just for openers. There would be appearances on the big weekly programs sponsored by Coca-Cola, Chase & Sanborn, Camel Tobacco, Lucky Strike, Palmolive, and Standard Oil. Of course, I was not the star of these shows. These shows all featured luminaries of stage and screen. Eddie Cantor, Jeanette MacDonald, the Boswell Sisters, Kate Smith, the Mills Brothers, and Burns and Allen are among names that come immediately to mind, for radio, like the Palace of the past, was now attracting the finest talent available.

Billed as an extra added attraction, "the world's youngest record holder" was on her way to a whole new line of work. But there was a decided change in attitudes at home. Not by Mother. The Daggett & Ramsdell idea delighted her. She felt that I would now be in a position to clear up many of the public's misconceptions about flying, and besides, she loved the company's cold cream.

But Father found my actions difficult to understand. In his view, appearing on a platform with Eddie Cantor or Ed Wynn was show business whether it took place in front of a microphone or on a stage. The fact that my appearances were spotted after station breaks so that I always spoke to the audience directly and never participated in the performers' part of the show did little to placate him. After all the fuss I'd made about not going into show business — *now* look at me!

I tried to tell him how much I wanted to become a part of the aviation world the altitude work had given me a glimpse of. My conversations with Doolittle, Williams, Ofstie, Miles, and the Soucek brothers had shown me how far away from the core of aviation research I really was. In 1930 the value of the physicist and aeronautical engineer was just beginning to be fully appreciated. Up to then it had been mostly trial and error with decisions being arrived at by way of pilot's experiences. Such things fed the imaginations of aircraft designers and helped the engineers but did little to open up the horizons promised by the advances already made through the Guggenheim grants, to such innovators as Pan American, Robert Goddard (the father of rocketry), and Lawrence Sperry's incomparable contributions to blind flying. My formal education was too meager even to allow me to be considered in this area, but the way the money was unexpectedly rolling in, I was still young enough to get my college degree. Then maybe...

Trying to make Father understand all this was difficult. I knew he would ask why I hadn't taken all this into consideration when he and Mother were so insistent on advancing my education, and I didn't have any answer to that. In 1927 I had barely heard of an atom, and now to learn that it was on the verge of being split — at least according to my navy friends — was indeed pie in the sky. Also, Father was still flying Wacos around, and my attempts to describe the pleasures of flying heavier craft were not too successful. I know now he was afraid for me. But at the time I thought he was uninterested in hearing about the Bellancas.

The Daggett & Ramsdell show was fun to do. With that and the big specials, it wasn't long before NBC's audiences had been introduced to my flying friends — Frank Hawks, Roscoe Turner, Clarence Chamberlin, G. M. Bellanca and Bert Acosta. The programs were enjoyable because my guests were an articulate and gregarious lot. Not once did I face Phillips Carlin's problem. We never used the scripts provided by the studio. They weren't accurate, so why bother? As long as we didn't say anything to land NBC in the courts (and we didn't), the executives looked the other way.

The mail reached unbelievable proportions

in a matter of weeks. It came largely from young people in my own age bracket, and I went through it with the aid of two secretaries and chose questions from it to answer on the air. All this was encouraged by NBC to build my reputation as a broadcasting personality. The announcements of attempted east-west Atlantic crossings were now starting to appear in the papers. They were hopeful that I was personally acquainted with the flying crews on these daring ventures. And I was. It would make my job much easier.

The west-east Atlantic crossings had been done enough times that only round-the-world or great distance endeavors would gain much attention. But east-west, Europe to America, had yet to be accomplished. There were many good reasons for this. Headwinds, fogs, and storms were the norm for this crossing in the teeth of the prevailing winds. It was, in fact, so bad on this route that there was later evidence that at least one crew of Atlantic fliers coming from England had actually made it to the U.S. before crashing in the wilds of Labrador, overflying the coast by many miles when they were unable to get their bearings.

Deke Lyman swore me to secrecy when Major Charles Kingsford-Smith, Australia's most famous flier, was signed by the *New York*

Times for the exclusive story of his east-west transatlantic attempt. He planned to do it in his three-engined Fokker monoplane, the same one which he had flown from California to Australia in 1928. Deke wasn't sure if he planned to make the flight from Holland or England, but it was definite that the *Southern Cross* was now in Europe and being readied for the long ocean voyage. I couldn't help praying he would be the first to make it. I'd met him briefly in California and been absolutely enchanted with his accent and his stories of flying "down under."

This was a side of broadcasting I approached warily. The thought of shoving a microphone into the face of an exhausted pilot and asking him how he felt after battling his way across one of the world's stormiest seas didn't really appeal to me. I *knew* how he felt — and asking him to talk about it for the edification of millions of unseen listeners just might elicit an answer that wouldn't pass the censors.

One thing that always annoyed me about free-lance flying was the way in which representatives of organizations like the American Veterans of Foreign Wars, the American Legion, the Red Cross, and even the Boy Scouts of America expected a free flight exhibition for their fund-raising air shows. I didn't mind doing it, although it cost time, money, and aircraft maintenance, until Jim Smith ferreted out the information that these shows were always run by paid promoters. In most cases they were supposed to pay all of us. If they could bluff us into doing it for nothing by insinuating our lack of "patriotism" if we didn't, they could (and did, I found) pocket our money, and the organization was none the wiser.

Dealing with NBC was something else. The network would negotiate directly only with the organization, not the promoter, and unless proof of the actual destination of the gate receipts was presented, the organization was graciously turned down, saying I was not permitted, under my contract, to appear. It was amazing how quickly this cut the freebies down to a minimum and how my income soared after that.

I was just settling in to this new and pleasant existence when a note from Russ Holderman brought me up short. He reminded me that I still didn't have my transport license and that it was something I could no longer afford to be without. Since I had all the qualifications, why wait until I was older? Actually I could have applied for it anytime after August 17 of the previous year, but I was dropping jumpers that day over Boston's airport and had kept busy ever since with one thing and another.

I really wanted that transport license. It was the most coveted piece of cardboard in a pilot's flight bag. Regarded as the pilot's doctoral degree, it was the ultimate proof of your competence to fly any aircraft anytime, anywhere. Also, I must admit to wanting to make G.P. squirm a bit. I knew he could stand for any female pilot in America getting that license but me.

I called Russ, and we made arrangements for me to take the test at his new airport located near Binghamton in Le Roy, New York. Long Island was just blossoming in the balmy weather of an early spring, but the Finger Lakes area was still in the icy clutch of winter. I sorely missed my Eskimo bunny suit, but by borrowing bits and pieces from Russ and his instructors, I made do nicely, although I refused to allow any cameras within a half mile range. As I've said, *nobody* could look attractive in a fur-lined flight suit, but when the suit, boots, and helmet didn't match or fit, the overall effect went swiftly from unbecoming to ludicrous. However, I was there to work, not worry about my appearance or make explanations to spectators.

I would be taking my flight tests in one of

Russ's student training planes, a little Fleet biplane. She was a sturdy, stable little craft that I needed only a few minutes of flight in to become completely familiar with her foibles and characteristics. I had no qualms about passing the flight test on merit, but I did dread the possibility of running into an official who was dead set against women pilots in general and eighteen-year-old ones in particular.

I had already learned to my sorrow that the stinging price of success often meant a sight-unseen dislike by individuals who felt a need to bridge the "distance" between us created by the headlines that followed in the wake of my activities. I prayed this would not happen here, and my prayers were answered.

When the inspector arrived, I found him to be a friend from the staff of the Roosevelt Field branch of the Department of Commerce, so all was well. Not because he was a friend, I hasten to add, but because he was tough, disciplined, and fair. He would never pass me if I failed, but he would never fail me if I passed — and you can't get any better treatment than that.

Owing to Russ's expertise, the Le Roy Airport was one of the most brilliantly designed private airports I have seen. It had been planned as an exclusive flying club, and its runways, landscaping, and architecture were years ahead of their time. For wealthy vacationers, the location was superb. But the deepening economic depression had already discouraged some of Russ's charter members, and it was clear that the airport would have to become more of a commercial enterprise than he had first envisioned.

I wanted very much to help him, and the only way I could think of was by my drawing national attention to the field. For the time being, that was impossible. Russ was adamant about keeping the press off the field until the tests were over, a stand which made me more grateful to him than ever. It gave me the freedom to hone my loops, figure eights, and precision spins in complete privacy, something that would not have been possible for me anywhere else in the country.

But I felt sure there must be a more exciting way to attract long-term investors than by having the story of my getting the transport license by-lined from Le Roy, New York, and topped by a picture of us with the department inspector. A day or two later I took the tests, and the inspector told Russ privately that I had passed. He wasn't supposed to do this. Usually there was an agonizing three weeks' wait for this verdict, which came by mail from Washington. But this license had never been issued to anyone in my age-group, and there was so much attention focused on this particular test, that he relaxed the regulations.

It was a banner day for me. I had an even more important reason for celebration. Mother had forwarded my mail that morning, and in it was the approved draft of my first article for *Liberty* magazine. Enclosed with the draft was a note from Fulton Oursler, *Liberty's* senior editor, congratulating me on my first literary effort and asking that my next piece discuss in detail "the current licensing procedures of the Department of Commerce for fledgling pilots." Talk about a made-to-order opening! I could hardly wait to get started on that one, which, in my mind's eye, already was heavily illustrated with pictures of the ideal airport (Le Roy) in which to take said tests.

My writing for *Liberty* magazine had come about in an interesting way, too, and points up one of the nicer aspects of those times. Important helping hands were extended from unexpected sources. I had appeared on the "Coca-Cola Program" with Grantland Rice, then the dean of American sportswriters. A true southern gentleman of the old school, he took me to lunch on the day of the program to outline how he wanted it to go. In the course of this conversation he asked if I had ever done any professional writing. When I said no, he said that was just fine because he was sure I would do very well at it. I will never know what gave him that impression, but it resulted in a luncheon with his friend Frank Crowninshield, publisher of *Vanity Fair* magazine. *Vanity Fair* was then one of the top "slick" picture magazines on the market. In each edition they featured a page of pictures of people worthy of being watched for progress in their professions. According to Rice, Crowninshield was thinking of me for this page. I was skeptical, but Rice was right, and I felt like Alice in Wonderland when the next issue of the magazine carried my picture and I found myself featured above Henry Luce.

What happened next left me even more dumfounded. Mr. Crowninshield called Fulton Oursler on *Liberty* magazine and told him that I would be an excellent aviation editor for them with my appeal to their younger readers. I forgot about being dumfounded and became frightened. I had no idea if I could write well enough for publication, let alone be an editor, but he and Rice soothed my fears, saying I must leave everything up to Mr. Oursler. I have no idea now why I thought a novice would be given no help on one of the most successful magazines of the day. I only recall that the whole thing scared me so much I almost didn't keep the appointment with Oursler. But Mr. Rice kept after me in his kindly fashion, so I knew I simply had to go.

I found Fulton Oursler rather forbidding at first, but since Bernarr Macfadden (his millionaire health-faddist publisher) was a flying enthusiast, I got the impression that Mr. Oursler was bowing to the inevitable, and if he had to have an aviation writer, it might be better to have a novice he could control.

I went back to Freeport, took the family typewriter up to my room — and dug in. He had said to write as if I were talking over the radio, so that's exactly what I did. He removed nothing but a few unnecessary exclamation points, and the piece went into print just as it was. Because of the timeliness of the material, the piece was well received, and I was hired. I sent the first copy I could get my hands on to C. B. Allen at the *World* with a note saying, "*See* — and you said I couldn't write my name!" I got one back almost immediately saying, "I could have sworn you'd have trouble with the Smith part — C.B."

On the same day I took my tests at Le Roy, one of Russ's pilots had asked me if I would take a local businessman up for his first flight. The businessman had been watching the tests and asked specifically for me to fly him. When we were introduced, I stared up at him, unbelievingly. He must have been six feet five inches and weighed in at at least 300 pounds. Getting him into the front seat of the Fleet trainer required extreme diplomacy and a couple of shoehorns.

Luckily it was a gusty day, and the wind lifted us off the ground more effortlessly than would have been possible under other conditions. Bravely the little Fleet climbed up over the rolling hills to give my passenger an excellent view of the snow-patched countryside. He seemed to be enjoying himself immensely, but he was so big that any movement of his to see over each side of the front cockpit blocked my front view completely.

When he signaled he'd had enough, I shoved the nose down and pulled the throttle back to land into the wind. The winds were now shifting so rapidly that the windsock was almost twirling around in a full circle. When I got down fairly close, I saw I would have to use the shortest runway, which meant an abrupt approach down the side of a steep hill. But I also saw that there was a graded hillock at the runway's end that would cut our forward speed, so I thought no more about it. I knew I should come in in a steep forward sideslip, flattening out and losing speed with a series of wide fishtails, but I didn't want to discourage my passenger's new interest in aeronautics with this kind of fancy footwork, so I opted for a steep downward glide. The wind shifted again and blew me right back up. Opening up the throttle, I found I practically had to fly the plane onto the ground. I had to sideslip a bit at the very last to make sure I didn't overshoot the runway, but everything was working out well because my passenger's weight was helping us descend. When I pulled the stick back to set down, we were suddenly ballooned upward about twenty feet by a wayward gust of wind that vanished as suddenly as it appeared and left us stranded there. Today this would be called a wind shear. In 1930 we called it a vacuum. But whatever you call it, the effect was as if a giant suction had suddenly removed every bit of air pressure from the control. Technically we were in a stall, and without air pressure I couldn't get out of it. Instinctively I slammed the throttle open and prayed. Two things happened. The wind came back as suddenly as it vanished, and the little Fleet set down as light as a feather at full throttle!

Putting an airplane on the ground at full

throttle can be described as a good trick only if you can get away with it, and I managed to do it that day because the runway was built on an incline. To make matters even more complicated, this particular Fleet was one of the trainers without brakes, but Russ's runways had large taxi circles at the ends so after running uphill and losing the last of my forward speed, we careened around the runway's arc and came to a halt. My passenger was so impressed with my skill that he asked if we could make another trip later in the day.

I thanked him for the compliment but begged off, citing a previous engagement. I forget now if I had one, but I know I wasn't about to tempt fate twice. As he was helping me out of the cockpit, Russ said casually, "I meant to tell you, Elinor, not to use that runway except in an emergency. We discovered that those trees create such turbulence that it's hard to make your approach over them, and that's why I had them bank up the end of it that way." *Now* he tells me!

The *Liberty* article duly appeared, and Russ was kind enough to say that the publicity was of great help to him. I hope so. It was such a beautiful airport, and he had worked so hard on it. I only wished I could have done more.

NBC got a lot of publicity mileage out of the

granting of the transport license, although I held the radio people off until I actually received it. One of the first to call and congratulate me was Amelia. She suggested we go to lunch the following day to go over some additional ground regarding the Ninety-Nines. I was still hanging back because of her insistence that the organization lobby for the inclusion of women pilots in the big money races. I simply could not envision the rank and file of the Ninety-Nines cutting pylons against Jimmy Doolittle, Al Williams, Jim Haizlip, and Roscoe Turner. But she made it very clear that she wanted me in the organization, and with that one reservation, I really wanted to be. Heath and Omlie were still balking, too, so I promised to talk it over with them as soon as I could to see if we could come up with an acceptable compromise.

Our orders arrived, and as we started to eat, she asked, "Would you believe me, Elinor, if I told you how often I have tried to get you and G.P. together again?"

"*Again!* Like that day in Chicago when he handed me that line about my qualifications measuring up to those mythical standards of his?"

She smiled wryly and nodded, saying ruefully, "Well, maybe he *did* get a little carried away, but when he told you that, he really meant it. He *was* planning that management thing — but it just didn't work out." (I didn't tell her that I'd heard from some of those interviewed that the reason Putnam's management scheme didn't work out was the meager amount of money offered to keep the pilots under contract and the hefty bites Putnam planned to take out of any incomes they might generate on their own.)

"Amelia, I don't doubt that at all. But why didn't he inform me of his change in plans before he tried to knock me out of the sky altogether?"

"He did *what?*"

I told her in detail about the confrontation of the year before.

"So you see, bringing us together would be an exercise in futility. It's not even a case of mutual dislike. On my part, it's contempt, and I make no secret of it."

"I hope you know I was completely unaware of any of this," she said anxiously. I nodded and said I thought as much. "But he'll do anything he thinks will protect my interests. I just don't know how else to apologize for him,' she finished lamely.

"I do sympathize with your position in this, Amelia. But 'protection of interests' and just plain bullying are two different things. He may have the rest of the world conned into looking upon him as a publishing tycoon, but in my book he's a slick press-agenting promoter. And just in case our paths don't cross again, you can tell him I said so."

"Well, that makes the other reason for this meeting more difficult than I anticipated. You have always had such good rapport with the New York press that I was about to ask you what you thought their reaction would be if I married him."

I looked at her in mock horror. "I think they'd as soon see you hooked up with Genghis Khan. Anyway, the last time I looked he was very much married — and wasn't he married once before that?"

She nodded. "My mother and sister are opposed for the same reason, but he and his wife are currently separated and will be divorcing as soon as the details can be worked out."

Privately I couldn't help wondering if Dorothy Binney Putnam was aware of all this. After about twenty years of marriage it must have come as a bit of a shock. It also occurred to me that this was a strange turnabout for Amelia, who had always publicly declared herself totally uninterested in the wedded state. On the other hand, maybe she, like Lady Mary, had discovered the going too tough without a millionaire — real or reputed — in the background.

In that respect it could be a marriage of convenience on both sides. Given G.P.'s extremely possessive nature, it would assure him that his prize could never get away. As for Amelia, her standing in aviation circles was not high at this particular time, and she wanted very much to become what he had been claiming. If he pulled away all support if she didn't marry him, she would be in a very bad way. By this time he'd made so many enemies in the aeronautical industry that I knew the job offers for her services were actually nonexistent. But he kept right on putting out bulletins and dreaming up flights for her to perform in order to keep her in the public eye. Unfortunately he still refused to face the fact that she didn't have enough air time and experience to do these things safely. Simply put, he'd made her a star before she'd learned how to act. But he *was* making a lot of money for her, and many of his personal contacts still held up well.

I knew how desperately she wanted to hone her skills, and I tried to make her promise that she wouldn't take on any more junkets without getting more serious and adequate instruction. She agreed, but it was half-hearted, for she said the schedule he'd already set up for her for the coming year barely left her time to wash her hair.

I was sitting facing the door. As we talked, a man named Fitzhugh Green came into the restaurant. Green was one of G.P.'s string of ghost writers and was well known to the flying fraternity. Amelia's back was to the door, and when I told her that Green had just come in, she paled slightly. When I told her that he had joined some friends, she seemed greatly relieved, and I noticed that she didn't stop at Green's table on the way out. Was she worried that he would tell G.P. she'd been lunching with me? I'll never know the answer to that, for it was the last time we dined together. From then on our paths diverged as widely as our outlooks on life.

Much as I loved flying, I never wanted or expected to have it dominate my whole life. She very definitely did. She wanted the acclaim that came her way, and she worked very hard to get it. She didn't particularly want children in her life, whereas I couldn't envision a future without them — after I'd attained my goal of being a top expert at the controls, that is. The importance of detailed preparation in every facet of the flying business had been pounded into me by experts, so I would never dream of flouting their teachings.

But many people — starting with her close friend Paul Mantz down to those members of the Coast Guard and navy who participated in the ground crewing of her transoceanic triumphs — reported Amelia's cavalier attitude toward vital detail. She was privately criticized for her lack of radio usage during the Atlantic flight and for her lack of actual knowledge of available radio techniques at the time of her disappearance in 1937. But she had a raw courage that carried her through, every time but the last one.

Looking back, I know now that I never had that kind of daring. I was more than willing to take on long-distance flying, whether transoceanic or continental, but I wanted everything in the way of technical help I could get. Two years after my luncheon with Amelia I had the backing and the airplane to make a nonstop flight from New York to Rome. But because of the deepening Depression, my sponsor had been forced to cut back on expenses to the point where the latest instruments would not be available to me. I weighed the pros and cons for two days and then canceled the flight. It was one of the most painful decisions I have ever made, but the odds against making a safe crossing of the Alps at night without an earth inductor compass were just too great in my mind. So I did the next best thing. I sold the Lockheed to Amelia, and she flew it from Honolulu to Oakland, California, the first time this had been done by a woman.

C. B. Allen became quite close to Amelia during the latter part of her career. He was with her in Miami at the start of her final flight and acted as liaison with the press there. According to C.B., the decision to leave before the arrival of the radio equipment that would have permitted her to communicate directly with the Coast Guard cutter *Itasca* off Howland Island was made for reasons different from those made public. Money was very tight in 1937, and G.P. had been forced to make financial commitments for her that would require a very tight flying schedule. I don't think he did this without serious concern for her safety, but she had had a crack-up in Honolulu that had put this planned flight back a good six months. Furthermore, the repairs to the twin-engined Lockheed had been horrendously expensive.

The crack-up occurred because Amelia had not listened to her flight engineer, Paul Mantz. Synchronizing throttles on two- and three-engined airplanes in those days was largely a matter of ear and instinct. But Mantz didn't leave anything to chance. He figured out the exact tachometer settings for her to follow on each engine for a fully loaded takeoff and warned her over and over against "jockeying" the throttles to "play it by ear." At best this was a tricky business to be done only by an expert, which she was not. She did move the throttles out of "sync," and the big bird careened off the runway, collapsing part of its landing gear under the weight of the gas load. The ship had to be dismantled, shipped back to the United States, and the whole flight re-planned in the opposite direction owing to seasonal weather conditions around the world.

In spite of this setback, she never once thought of canceling the flight. It probably made her more determined than ever. G.P. worked very hard to keep ahead of the bills, but he was jealous of her friendship with Mantz and kept chivvying him on payment. This, in my judgement, was one of G.P.'s most serious character flaws. He was firmly convinced that his exalted social position gave him ample authority to cheat the peasants whenever he could get away with it. But Mantz was no peasant. He was an excellent businessman who had formed the Hollywood stunt pilots into a well-paid group with himself at the helm, and he was also one of the best pilots in the United States. G.P.'s constant attempts to humiliate him drove Mantz to sever all connections with the flight.

I am convinced that had Mantz been on the scene when Amelia left Oakland for Florida, the parachutes would have been on board, not left in the hangar. Amelia told C. B. Allen that she didn't want to bother with their weight, and besides, they'd be of no help to her over the water. It was on the tip of C.B.'s tongue to tell her how invaluable they might prove as shelter from insects, should she come down in a jungle, but he told me she was in such a strange frame of mind that he thought better of it. She could be very stubborn at times, and it is my personal opinion that this characteristic surfaced each time she reviewed in her mind what she'd gotten away with on previous flights and decided to gamble on taking the same chances over again.

Naturally I'm not stating this as a fact. Nobody can. But it certainly is a logical explanation for the last wild chance she took by leaving without the radio equipment. C.B. said she told him she was taking off because G.P.'s nagging was driving her up the wall, and she didn't want to wait any longer for delivery of the equipment. For once in his life I don't think he was nagging. I think G.P. was terrified of financial disaster. Any more delays would completely throw off the schedule of the personal appearance tour that was planned for her return, and he simply couldn't afford numerous cancellations. By that time everyone's nerves were snapping, and Mantz was no longer around as pilot, adviser, and overall major-domo. Putnam was trying to fill those shoes, and they simply didn't fit.

However, all this sadness was still seven years into the future on that day in 1930. I really hated to see Amelia marry G.P., not only because of my personal dislike but because at no time during our conversation did I get the impression that we were discussing the romance of the century, and I, of course, was a hopeless romantic.

Somewhere in the back of my mind I was sure there was somebody out there who would propose to me because of what I *was*, not what I did — and that was the man I was holding out for. Of course, I entertained these serious thoughts only on Tuesdays and Fridays. The rest of the week I was much too busy working or just completely enjoying being a successful aviatrix who had a *transport license.*

A great deal of my euphoria could be traced to NBC, for it did everything possible to coordinate my flying with the radio broadcasting.

Thanks to the additional recognition provided by radio, requests for personal appearances soared. I was more grateful than ever to Red for the grilling training he had given me, as I found myself flying out of race track ovals and ball parks and – once – landing on the edge of a pier in Ocean City, New Jersey, after an aerobatic display.

Tim Sullivan, now a program executive at NBC, was exactly like G.P. in one respect. He never learned the limitations of the airplane. If an organization was legitimate and checked out financially, he was inclined to overlook the fact that their town didn't have an airport and they would like me to make a landing in a nearby ball park, provided they could get the players off it in time. I *really* balked, though, when a group in West Virginia wanted me to land on top of the new high school gymnasium and Tim assured them it would be "no problem." Maybe not to him, I decided, but this was where I got off. I did suggest that he call Frank Clarke, however. Told him it was right up Frank's street!

I greatly enjoyed broadcasting and still feel it was the easiest money I ever earned. I always had plenty of material for my thrice-weekly show just by reading through my mail. The big shows sponsored by Coca-Cola, Lucky Strike,

Camel, Standard Oil, Colgate, Palmolive, among others were even more fun because they gave me the opportunity to meet some delightful performers. Jeanette MacDonald, opera star Reinald Werrenrath, Charles Previn (uncle to André Previn), Lennie Hayton (Lena Horne's husband and MGM's musical conductor mainstay), radio star Bill Hughes, movie star Bebe Daniels, and light opera star James Melton are just a few of those I worked with. This may seem like a strange programming mix, but flying was still of such compelling interest to the public that it seemed quite normal at the time.

It would be interesting to report fits of temperament on the part of some of radio's stars, but I saw very little of that. In fact, the announcers were far more skittish and jealous of one another than the performers, to my surprise.

I didn't have this experience with Floyd Gibbons, though. Gibbons was then an extremely popular newscaster. His delivery was a rapid, machine-gun type of patter, completely new to the airwaves. I had heard from reporters how highly competitive he was in the news field and how jealously he guarded his success. Maybe to his competitors he was a hard man to deal with, but I found him an absolute pussycat. Old enough to be my father, he took me under his wing, giving me valuable tips on diction, program presentation, even network politics. The only word of criticism I ever heard pass his lips was directed against an announcer named Graham McNamee. He said the same things about McNamee that I'd heard about him. As far as I could see, McNamee was a loner, but I had to work with him only once to find out why.

We had been sent to Washington, D.C., to broadcast the annual Navy Air Races at Anacostia. It was planned as a dry-run for me prior to my broadcasting the National Air Races, which were scheduled to take place in Chicago that fall. Since I was acquainted with most of the participating pilots, my job was to identify each of them, filling the listeners in on their backgrounds, as well as keep track of their positions in the race. As the races progressed, I was to announce their speeds as this information was handed up to me by the timers on the ground below our elevated booth.

It was my first time on a remote broadcast, and I was very anxious that it go smoothly. I waited for McNamee to turn over the microphone after he finished making the station identifications and setting the scene, but it quickly became obvious that he was off and running on his own. Lacking information didn't bother him in the least. His mellifluous tones and years of experience carried him right along. But I could see my big chance to broadcast the National Air Races going right down the drain if I didn't get my hands on that microphone.

McNamee was at the peak of his fame as an announcer, so getting the microphone away from him if he didn't choose to relinquish it was going to be the trick of the week. But the fates were with me. For probably the first time in his career he was caught short by a sneeze (airport dust), and he was forced to hand the microphone over as he muffled the sound in his handkerchief.

Each time we had a station break, I would hand the mike back, and at the end of each identification he would start talking about the crowds, the weather — anything at all but the race itself — as the airplanes kept right on roaring around the pylons. The sound engineer finally left his post, stationed himself at McNamee's elbow, and, the minute the identification was over, deftly took the mike and handed it to me. McNamee was trapped. The engineer gave him a reassuring smile but didn't move. It certainly looked as though he were operating under instructions, and I could see Mac's inner struggle. He couldn't *say* anything. We were on the air, live. Tape was still a distant gleam in an unknown technician's eye. But his demeanor was one of such contained fury that I knew the stage was set for fireworks the minute we went off the air.

I had specifically requested permission from the navy to interview Lieutenant Alford J. Williams, the navy's top aerobatic and speed ace and a good friend of mine. He was knowledgeable and articulate, and I had deliberately picked him as the star of this show. But McNamee's proprietary air as he draped an arm around Al's shoulders quickly told me who was actually going to conduct this interview. I was shocked and unhappy. I wanted Williams to *talk*, and I knew full well that when Mac got his hands on that mike, this would be one of those second-person "and then you did" discussions at which the honored guest sits there, dumbly contributing an occasional yes or no.

But the sound engineer knew Mac's tricks. The minute the station break was over, Al and I found ourselves facing each other, talking into two new microphones that had been quietly placed in front of us. Mac's was disconnected, and he was off the air.

The program went smoothly ahead. Al was even better than I had hoped. He had such a rich background in international racing (he was one of the U.S. representatives in the Schneider Cup trials in Europe) that I had only to prime him with a few pertinent questions and sit back and listen. When it was over, we chatted a few minutes, and I looked around for McNamee, but he was nowhere to be seen. The sound engineer pulled me aside and in a worried whisper conveyed the news that the minute we went off the air, McNamee stalked out of the booth, saying he was going right back to New York to get this little lady "straightened out."

I had flown a fast clipped-wing Waco down to Washington, and it seemed like a good idea to see how fast she could make it back to New York. If the eggs were about to hit the fan anyway, it would do no harm to see where the shells landed. With a reassuring (which I didn't truly feel) wave to the loyal engineer, I left the booth quickly, warmed up the Waco, and took off. I flew straight back to North Beach Airport and took a cab into New York, beating Mr. McNamee quite handily to Aylesworth's office.

I must admit that I feared for the engineer's job as well as my own, but I did not want to be the one to open up the discussion. I wanted that to come from McNamee. There was always the possibility that the network had placed me in a no-win situation by sending a man of his stature on their first broadcast like this and not telling him fully what they expected of me. Fortunately the broadcast was such a departure from anything NBC had ever done in this area that I had a perfect excuse to lead Mr. Aylesworth into telling me what he thought of it.

We were interrupted by a long-distance phone call for almost thirty minutes. He apologized as he hung up and said the program was very well received and that they were already getting excellent responses (phone calls) from the minute it went off the air. He wanted to know if there were any other races worth giving this same attention to, but I pointed out that local racing would have only local appeal. However, according to the *Times* story of the day before, the Australian ace Major Charles Kingsford-Smith was about to try the east-west crossing of the Atlantic in the very near future. If he managed it, it would be the first time this had been successfully accomplished.

This kind of derring-do appealed strongly to Mr. Aylesworth's sense of the dramatic, and an hour later I was still holding forth on Kingsford-Smith's background – World War I ace, first man to fly the Pacific from Oakland to Australia, back in 1928, his plans to make this

391

Atlantic crossing part of the first round-the-world flight that included nonstop crossings of such wide ocean spans; etc. – when in strode Mr. McNamee.

Seeing me there, he turned beet red. "How did *you* get here?" he bawled.

"Flew," I said quietly.

McNamee launched into a tirade against the engineer for handing over the mike to me – and last, but not least, against little Miss Flier, who had loused up his interview with Al Williams.

"Your interview?" asked Aylesworth. "Wherever did you get that idea? Miss Smith was the one who set up the interview – not our news department. Who said you would interview Williams?" McNamee's demeanor changed as he sought to retreat with some semblance of dignity.

"Well, I – uh, assumed – I always do the key interviews, and nobody said I wasn't to do this one ..." His voice trailed off lamely.

"Hmm, well," said Aylesworth briskly, "I suggest you don't make any more assumptions in the aviation area from now on." It was pleasant to have won this skirmish, but by the expression on McNamee's face I knew this war was far from over, and I hadn't won *that* – not by a long shot.

I waited until he left to gain Mr. Aylesworth's ear regarding the sound engineer in Washington. He assured me that I need have no fear of reprisals.

I met Floyd Gibbons as I was leaving the building. "Been flying?" he asked, noting my outfit.

"I have indeed. I have also done my first remote at the Anacostia races."

"Great . . . how did it go?"

I told him briefly, and he cursed under his breath. "That lousy b———. He knew perfectly well he wasn't supposed to take over that show, but he thought because you're a kid, he'd bluff you out of it."

"Well, he didn't get away with it because of two things. You alerted me to be wary, and the sound engineer took care of everything else."

"Yes, this time, but there's a question in my mind. I'm afraid that if you hadn't beaten him to Aylesworth's ear, he would have shut you out completely."

"We'll never know, but of course, I don't want to think so. After all, I said nothing about any of it before he got there. He just barged in and aired his grievances, which Aylesworth didn't buy."

"Yeah — well, keep a weather eye on him just the same, and don't trust him while he's standing alongside you!"

As I told Mother about it that night, she listened intently and said, "Sis, have you looked at this from McNamee's side? He's at the top of a very competitive business. He's not a young man, and he considers anyone like you a distinct threat. In this instance not only were you broadcasting the races, but you set up the interview with one of the navy's most important star pilots. McNamee knew what an attraction he is and what stature you gained for the network by producing him. No wonder he was angry. More than that, by the time he got back to New York, he'd had time to realize that he was not in a position to do any of that and was probably more desperate than angry."

"If he was, he fooled both Aylesworth and me."

"Maybe you, but not Aylesworth. McNamee's territory was kept firmly at its present border. He will announce the big shows exactly as he has in the past, but he is not to impose himself in the aviation area, right?"

"That's what I came away with. I just wish none of it had happened."

"Of course you do, but this time it couldn't be helped. Try to learn from it and avoid any of these situations in the future. You well know how bitter show business feuds can become, and this could easily take on that form. Floyd Gibbons means well, but he's fighting McNamee on another front. Make sure you don't become embroiled in that."

"You're not recommending that I turn away from Floyd, are you? I have already suggested to Mr. Aylesworth that he would be an excellent co-announcer for the Kingsford-Smith landing. I want him because he did a lot of World War One reporting and Kingsford-Smith flew for the Allies."

"I hope you made the suggestion before McNamee arrived and staged his scene."

"Luckily that's how it was. I had no guarantee that he'd really come straight to Aylesworth's office, and I certainly didn't want to do any fussing if he had changed his mind en route. But I really don't think there will be any trouble. Floyd has such a colorful and authentic war record that he's the logical choice for this job."

In a matter of weeks all the newspapers were ablaze with Kingsford-Smith's plan to fly the Atlantic in the *Southern Cross*. Signed to write the flight's story exclusively for the *Times*, he was doing everything he could to avoid the European press, driving them to making up their own stories, and speculations on his takeoff point ranged from Paris, France, to the Outer Hebrides. All during the month of June public interest in the flight was building to a pitch of excitement that rivaled that of Lindbergh's crossing three years before.

There was no question that it was a more difficult flight. Kingsford-Smith would be bucking all the way the same winds that had been so kind to Lindbergh. Where Lindy had a thirty-mile tailwind a good part of the way, Kingsford-Smith knew he would encounter thirty-mile-an-hour headwinds. On the other hand, radio had made tremendous advances in three years' time, and radio people would be flying the whole distance with two trailing antennas in order to talk to the shipping lanes and directly to the *New York Times*. Instead of one man's pitting himself against the elements, there would be four men aboard a three-engined monoplane. But their goal of being the first to make it a nonstop trip would try their endurance, for the fogs over Newfoundland were known to roll up to unusual heights, and if they flew too high in getting them over this obstacle, there was the very real danger of ice forming on the wings, which, in 1930, there was no way of preventing.

Since 1927 numerous pilots had been lost on

this east-west attempt, starting with Charles Nungesser and François Coli, who left France a few days before Lindbergh took off from Roosevelt Field. They were never sighted or heard from again. Shortly after this, Captain Leslie Hamilton, Colonel F. F. Minchin, and Princess Anne Lowenstein-Wertheim took off from Uphaven, England. It was their plane that was tentatively identified from wreckage discovered by trappers deep in the wilds of Labrador some thirty years later.

In August 1928 Captain Frank Courtney and three crewmen attempted to fly westward over the Atlantic from the Azores in a Dornier Wal flying boat. They were forced down at sea after battling headwinds for fourteen hours. They were rescued, but the flight was abandoned. Miss Elsie Mackay and Captain Walter Hinchcliffe left Cranwell, England, during that same year, heading for the United States. Their plane was sighted in the water by a ship 170 miles off the Irish coast, but by the time the rescue vessel arrived on the spot the aircraft had vanished, and no trace of it was ever found. In 1929 Majors Kasimir Kubala and Leon Idzikowski tried to fly from Paris to New York but were forced down in the Azores, where Major Idzikowski was killed when they cracked up on landing.

Except for dirigibles, the only westward flight to achieve even partial success was that of the *Bremen* in 1928. The *Bremen* was an all-metal low-winged monoplane built by the Junkers Corporation in Germany. Carrying Baron von Huenefeld, Hermann Koehl, and Colonel James C. Fitzmaurice, it took off from Baldonnel, Ireland, and was forced down thirty-four hours later, far off its course on Green Island, Labrador. It was such a bleak, desolate area that they were stranded there for several weeks before they could be rescued.

Kingsford-Smith was a perfectionist who believed in adequate preparation and total coordination of men and machines. His crew was hand-picked, and the *Southern Cross* was being completely overhauled at the Fokker plant in Amsterdam. To the press he said, "Nothing will be left to chance. So many things over which we have no control could go wrong that we are taking no liberties with those we can."

He refused to reveal his takeoff point, saying that he and his crews would be much too busy to give interviews or pose for pictures. I'm sure he was badgered into that statement, for reporters in those times could sometimes become maddeningly persistent. Also, Kingsford-Smith was one of the most even-tempered celebrities of the day. But it made for some nasty sniping from disappointed scribes.

He could not allow this to attain too much importance, however, and on June 4, 1930, the *Southern Cross* touched down on Baldonnel Airport, Dublin, to await favorable weather. They waited for an agonizing seventeen days before getting the go-ahead. On June 23 he and his crew left for Port Marnock, 7½ miles northeast of Dublin, where a 3-mile stretch of hard-packed white sand afforded better conditions for the heavily loaded *Southern Cross*'s final takeoff. Her weight when loaded was seven tons, almost half of which was gasoline. Her capacity was 1,300 gallons, but here he hedged his bet and took on 1,298.

"No use borrowing trouble with unlucky thirteen," he told the spectators. And when one of them asked if he felt any apprehension over losing his life in this venture, he replied crisply, "I have no wish to lose my life, and for your information, I have absolutely no intention of doing so!"

With that, he climbed aboard the big bird, motioned the mechanics to remove the planks they'd used to raise her out of the previous night's high tide, and with a jaunty wave opened up the throttles and was on his way to America.

13

Kingsford-Smith
Wears Sneakers, Too

Everyone was glued to the radio for the next day and a half. The *Southern Cross* had not been sighted by any ships since it passed over some trawlers off the Irish coast, but her radio was chatting cheerily with the world, so all was well. Originally cruising at 100 miles an hour, it was soon cut back to 80 and then 70 mph by headwinds. Kingsford-Smith told me later that he knew by mid-ocean he wasn't going to be able to make it nonstop, particularly if he hit fog banks off Cape Race. These formidable masses are caused by the warm waters of the Gulf Stream colliding with the frigid air swirling down from the Arctic and were always the greatest objection to taking the Great Circle route. But it was still the shortest distance over open sea for a land plane and for that reason had to be considered the safest.

By the afternoon of the first day out from Ireland the fog was so thick the *Southern Cross* was skimming the whitecaps. During this time her radio went out several times. As Kingsford-Smith said later, "That could not go on, for it meant that our antennas were dragging in the water. We were in danger not only of losing our antennas but of actually swamping the aircraft itself." He pulled up into the scud and flew blind for many hours.

By the following afternoon he and his crew were sure they were over Newfoundland but were reluctant to come down through the fog without getting a true bearing. Captain J. Patrick Saul, the navigator, was finally able to fix their true position and gave Kingsford-Smith the dismaying news that they were not even close to Newfoundland but were still more than 200 miles at sea!

This meant that the headwinds had been even stronger than they had thought, and they would have to abandon all hope of making the flight a nonstop one to New York. They would have to land and take on more gas at Harbour Grace, Newfoundland. But even that compromise was made more difficult when their instruments started to dance and flutter, due to unknown magnetic forces, and the additional static this caused on radio interfered with their reception. By now they were really worried that their gas would run out as they milled around, uncertainly trying to find Harbour Grace. At one point they climbed above the fog and called for a plane to come through it and lead them down. But before this could be arranged, the clouds suddenly parted, and they found themselves directly over the roofs of the city.

By this time the pitch of public excitement in the United States was building to an unbelievable peak. Once they had landed at Harbour Grace, it was a foregone conclusion they'd make it to New York. Even if it wasn't going to be the nonstop crossing the little Australian major had first envisioned, it was still the first successful east-west crossing and, as such, was an aeronautical triumph. The city of New York was approaching hysteria in its determination to honor these brave men.

I was busily compiling notes during the whole period of the flight. It would be my first experience on a remote-control broadcast of international interest, and I didn't want to find myself with any conversational dry spells while facing the broad back of a policeman or staring up into an empty sky, waiting for the *Southern Cross* to appear. It seemed to me that the listening public would want to know as much as possible about the backgrounds of the major's crew members.

I didn't get too much on the copilot, a twenty-three-year-old Dutchman named Evert Van Dyck with over 5,000 flying hours to his credit. But the navigator more than made up for it. It turned out that Captain J. Patrick Saul was a seaman, who had gained his master's license in sail and steam. The French government had awarded him the Croix de Guerre for bravery during the First World War, after which he had returned to the sea until Kingsford-Smith persuaded him to try another element. John Stannage, the radio operator, was a South African, educated in New Zealand and England. When Kingsford-Smith was lost during an attempt to circle Australia by air, Stannage volunteered as radio operator on a rescue plane sent out from Canberra after all hope of finding him alive had been abandoned. After three weeks of intensive search he found the major, who by then had been without food for fourteen days. Selecting Stannage for this venture went beyond being an expression of appreciation. Kingsford-Smith said many times over that Stannage's work on the radio was one of the key reasons for the *Southern Cross*'s success.

I was very disappointed to find that Floyd Gibbons was unable to work with me on the broadcast of the landing because of previous speaking engagements, but I held onto the hope that we would work together at some future time. Instead of Gibbons, I found myself paired with Curt Peterson, a young announcer with whom I had worked on some studio shows. If I couldn't have Gibbons, there was no one I'd rather have had than Curt. He was tall, handsome, amiable, and competent, and I knew I could count on him to roll with the punches if things went awry.

As reports on road conditions to the airports drifted in, I began to realize the importance of the police in handling unruly crowds. A call to Bill Deegan in the mayor's office solved that problem. It had been rumored repeatedly that the landing would be made at the new Curtiss Airport in Valley Stream, but Bill told me that this was a ploy to divert the public and that the actual landing would take place at Roosevelt Field.

Naturally I was delighted at this news. It meant I would be among friends, for Roosevelt was the home of the permanent press corps, now headed by John Frogge of the *New York Times.* John was originally of the same opinion regarding my flying as Carl Allen – highly skeptical. But as time rolled on and he saw that I was serious, he became a staunch friend and ally. If Kingsford-Smith harbored any reservations about talking to me because of his *Times* contract, I felt sure I could rely on Frogge to clarify the situation.

Curt and I decided that if at all possible, I would interview Kingsford-Smith inside the *Southern Cross.* We were banking on the crowds making it impossible for him to disembark. Curt would take up a position on a hangar roof prior to the landing, in order to describe the crowds below and any air traffic above. He would cut back and forth to me on the ground. Because of my familiarity with the field and its prevailing winds, I planned to station myself at the edge of the southwest landing strip. It would give me a clear view of his touchdown and a good idea of where he planned to taxi in. I got very nervous as the day wore on and the radio kept asking people not to go to the Long Island airports, while simultaneously reporting on the choked-up roads.

I called Deegan back and asked if he planned to send out any police. The Nassau County force would be totally unable to cope with a situation of this magnitude. He assured me that New York's "boys in blue" were already on their way with personal instructions from the mayor to help the "little lady in the white helmet with a microphone in her hand." Now all I had to do was lay hands on a white helmet.

I decided to fly into the field, rather than drive. I'd be surer of getting there because Curt had long since left with the sound truck.

I phoned ahead, or at least I tried to call North Beach, but the wires were jammed at both hangars. By the time I got out there I was surprised to find the field deserted except for Father's old friend Emil Burgin. He explained that everything that could even stagger into the air had been chartered by the press to fly north and meet the *Southern Cross.* Luckily for me, his own Travel Air was still available, *and* he was wearing a white cotton helmet. I promptly appropriated both of them and took off.

Upon landing at Roosevelt, I found that the traffic was building up so heavily that I had the Travel Air rolled into the Waco hangar. The past year had provided me with some painful experiences regarding the actions of a mindless mob, and I wasn't about to sacrifice Emil's airplane to them. Curt was already there, so we went into an immediate huddle. It would never do to get our signals crossed. CBS was nowhere to be seen. The *Times* had reported that Harry Von Zell would be its announcer, leading me to believe that the pressure must be off Husing.

As the *Southern Cross* flew down the east coast in leisurely fashion, Captain Stannage was busy at his radio key. By the time the big Fokker headed across Long Island Sound everyone in the metropolitan area knew where they were and what was happening. To make sure there would be no confusion about where the landing would take place, two escort ships took off early in the day, hoping to intercept Kingsford-Smith at Old Orchard, Maine. One had "ROOSEVELT FIELD" painted in huge letters on both sides of its fuselage, and the other carried cameramen and reporters.

We later learned that Kingsford-Smith had made one detour on the way to New York. Bernt Balchen was being honored by the town of Lynn, Massachusetts, for his work with the Byrd polar expedition, so his good friend Charles Kingsford-Smith and the accompanying planes flew over the celebration and dipped their wings in salute.

The *Southern Cross* was first sighted over Long Island at Port Washington on the north shore. There she made a leisurely right turn, ignoring the frantic signals from the Roosevelt

Field escort. Flying due west, she circled the city of New York, whose rooftops were dark with a cheering populace.

Curt and I had been on the air for some hours by now, and despite my copious notes, I was running short of material. Not to worry. Shortly after 7:00 P.M. I spotted the plane approaching us in the deepening summer dusk, her blue fuselage and pale wings gleaming in the last of the sun's rays.

I had made friends with a burly young policeman during the long wait. As the *Southern Cross*'s wheels touched the earth, he picked me up and shouldered his way to the spot I'd shown him earlier. The crowd surged through the police lines on the field, and Kingsford-Smith was obliged to cut his engines. Luck was with me all the way, for Van Dyck hopped down out of the cockpit, intending to help the ground crew. My policeman friend jumped up on the wheel and shoved me into the cockpit next to a very surprised Australian pilot. In spite of the noise and confusion, he remembered me from our meeting in California, so there was no necessity for any banal introductory chitchat. We couldn't get out of the cockpit for a good ten minutes until the police managed to restore some kind of order. It was a scoop beyond the wildest dreams of any reporter on the field.

I glanced down at the major's feet and saw that he was wearing sneakers. When asked about it, he laughed and said, "Always do. Over water, only sensible thing." Truly a man after my own heart! I pointed to my own feet, and we shouted with laughter. This victory was made all the sweeter by the presence of Ted Husing, whom CBS had obviously kept under wraps until this very minute and who tried to elbow my Irish cop friend out of the way. Ted stayed at the cockpit window, glaring at me, but it was obvious that even if I wanted to get out (and I didn't), there was no way I could manage it. The *Southern Cross* was a flying gas tank, and I quickly realized that Saul and Stannage, seated in back of the tanks, could never have changed places or directly conversed with Van Dyck and Kingsford-Smith at any time during the flight. This led us into a whole new conversational range, with Ted getting more apoplectic by the minute.

When we were finally able to disembark, Husing was swept back into the crowd. We later learned that even if he had gotten into the cockpit first, it wouldn't have helped him. CBS's field equipment broke down, and not one word of its broadcast reached the listeners.

I waved to Curt over the heads of the crowd,

and he streaked for the airport office, where he acquitted himself admirably. Neither the British nor Australian government officials had been able to get close enough to the *Southern Cross* to greet the crew, so they now took over before any more gaffes occurred. Before anyone was aware of what was happening, they had bundled the four airmen into the middle of a twenty-car entourage and headed for New York City, full speed ahead.

It was my first brush with the open warfare of news competition, and I must confess that it felt very good to win. Seeing Husing's determination to beat me to the draw made me realize that for these men the gathering of news was no game. I even saw some of my friends on the Roosevelt Field press eyeing me warily, and that *did* give me pause. Most of these men had been very good to me. Remembering Mother's words about McNamee, I talked with some of them, trying to make it clear that my foray into the news end of broadcasting would probably be brief. I was sure in my own mind that from the network's viewpoint I was a novelty that would wear thin once the business of flying settled down to being accepted as a legitimate way to travel.

Driving back home that night, I indulged in a mental review of the day's happenings, and it strengthened my resolve to stick even more closely to my chosen business of flying. I enjoyed broadcasting and was appreciative of the financial reward, but aviation was where I belonged and planned to stay, even if I had to do it for nothing.

The radio critics singled out Curt and me, giving us excellent reviews for the job we did on the *Southern Cross* landing. This was particularly gratifying since it locked in the National Air Races for me, and I was assigned one of the most important announcers in the Midwest to work with, Wally Butterworth. I'd never met Butterworth, but Floyd gave him high marks, and that was good enough for me.

During July and August, long after Kingsford-Smith had left New York to wend his way west, the bulletins started coming in again from Europe regarding another nonstop flight attempt – this time from France. Maurice Bellonte and Dieudonné Coste were poised and ready. But the weather stayed bad over the Atlantic for that whole summer, and I found myself facing a possible time conflict. The air races in Chicago took place during the last week in August, continuing through the Labor Day weekend. I had already agreed to fly a clipped-wing KR biplane there for the Aviation Corporation, making two exhibition ap-

411

pearances en route. When these arrangements were made, it appeared to be an ideal schedule. One appearance was at the New York State Fair at Syracuse for very good money, even though I found myself flying out of a race track oval again. I consoled myself with the thought that the city of Toledo boasted an airport, and that boded well for the success of the American Legion Air Show, at which I was to be the star attraction.

The enthusiasm with which I was greeted on both occasions brought home to me once more the power of radio. It was now five months since I'd set the altitude record, but the crowds greeted me as if it were yesterday. Coming into their living rooms made all the difference.

The FAI had still not accorded that altitude record official status, even though the FAI officials agreed that all of their terms and conditions had been met at the time. This was becoming an increasing annoyance because it denied the United States its true position on the international scene. The federation was slow enough to accord our male pilots' records, but the American women's category moved toward recognition with the speed of an Alaskan ice-age glacier. As nearly as we could figure out, recognition was being delayed deliberately to give the European female contingent an opportunity to pass the American women by. It was unfair all around since these women were granted access to military aircraft by their governments while the United States official stand was still *"no* females." Worrying about what was and was not fair is always a waste of time, but it bothered me that I still hadn't found a way to force the FAI hand. There was no doubt in the minds of the National Aeronautic Association – which calibrated the barograph – that I had, indeed, set another world's record. It appeared that this, along with the approval of Giuseppe M. Bellanca, would have to do for now.

Actually I had more immediate concerns to absorb my attention, for the dispatches from Paris were heating up. Coste and Bellonte were definitely going to make their try sometime before September 15. I prayed they would hold off that long, so I could finish up the National Air Races and have ample time for a return flight to New York. If the weather was favorable, I could easily make it in one day, but if it wasn't, I could be reduced to hedge-hopping to make it back in time.

The ship I was flying to Chicago was one I had been using all summer with great success. As a racer and aerobatic performer the Kreider-Reisner Challenger had no real competition in her class and weight. She was fast, agile, and sturdy; the only thing she wasn't was heavy. This worried me because September was the month for heavy thunderstorms and small tornadoes in the Great Lakes area over which most of my route lay. I was not anxious to take a drubbing as their unpredictable turbulence bounced the little ship around, nor did I relish the thought of the accompanying cockpitful of water that was sure to go hand in hand with nature's own brand of fireworks. I thought back longingly to the year before, when I'd ridden out those same storms in the dry comfort of the Bellanca's roomy cabin.

Finding the new Curtiss-Reynolds Airport northwest of the city of Chicago was easy. See that spiraling cloud of dust over there? Well, right under it is sure to be hangars, grandstands, airplanes, and pilots. The day was blistering hot – and so was my reception by Wally Butterworth. Having already decided that he was not about to like or give ground to this invader of his territory from the East, he made it crystal clear that he and he alone was in charge. For once I was too hot and tired to take umbrage at his attitude. Besides, I was only too glad to let him be in charge if that's what made him happy. I must have nodded in all the right places during this fiery monologue, for I was startled to hear him conclude his tirade with: "Say, where are you staying?"

I'd been booked into one of the downtown hotels some fifteen miles distant, and I gave him the name, assuming he wanted to know in order to be able to contact me in an emergency. As I turned to leave the hangar, he called out, "Wait – how'd you like a suite at the Edgewater Beach Hotel? I can get it for you for the price of a room." His ESP must have been twanging like a guitar string.

Chicago was a town I knew very well, having visited it many times with my parents. It was a great theatrical center, and its restaurants were of the best. Highest of all on the list was Henrici's, whose fine foods and hospitality attracted me like a homing pigeon whenever I was in town. But I knew how cruelly punishing the August heat of the Midwest can be, and I dreaded the thought of staying downtown. From the first moment that I glimpsed the airport's rising dust over the cockpit rim, I'd been thinking longingly of the Edgewater Beach. Far out on the lakefront, north of the city, it faced breeze-swept Lake Michigan. Did I want to stay there, far from the madding crowds and baking streets? Is the Pope Catholic?

Since the Fairchild Company was picking up my traveling expenses, Butterworth's comment about the price was most welcome, for I knew that Sherm would never spring for the price of a suite. (Incidentally I never called him Sherm to his face. I don't think anybody did.) Sherman Fairchild was the East Coast's Howard Hughes, and just as rich. Like Hughes, he had inherited his fortune, but unlike Hughes, he reinvested it and ran his own aircraft operation without benefit of any ongoing golden egg, such as that Hughes had in the Hughes Tool Company. Sherm's losses were handled on the spot, so where the dollar was concerned, you might say he was a bit snug. In fact, most of us called him downright tight, but that was probably because we would never understand the management of a fortune such as his. In fact, we'd never understand handling the *interest* on a fortune such as his. Be that as it may, Sherm had one outstanding characteristic that I admired. He stuck to his agreements. This meant that you'd better make a good deal when you went to work for him, for whether your contract was verbal or on paper, whatever you went in with was what you were going to come out with. In hindsight, I must emphasize his honesty in our own dealings despite our fairly numerous disagreements.

For Sherm, like most wealthy men, found it difficult to deal with people who felt they should be paid for displaying their talents. In his view, since he furnished the airplane for you to fly, why should he have to pay you to fly it? If his attitude seems insensitive, I must point out that he was not alone in this. Many other businessmen in aviation adopted the same posture. Strangely, many pilots accepted this in humble fashion, but I was not among them. I could always remember those long hours of shooting landings and stalling out on my back while I perfected my skills. Now that I had attained professional status, I saw no reason to advertise anyone's airplane without being compensated for it. But Sherm, used to dealing only with subservient secretaries and occasional PR writers, never got used to my obstreperous attitude, although he always paid up in the end. Strangely, he was upset usually by little things — the cost of gas and oil or, as I knew he would be in this case, the cost of living accommodations. His administrative assistant, Charles Robinson, and I had many a cup of coffee as we discussed these idiosyncrasies of his boss's. Robbie would beg me to downplay my independence. I knew he was right, but I simply couldn't crawl.

"Look, Robbie, I just don't have to. There are at least three other manufacturers in the area that will pay me almost twice as much for the radio and aerial advertising he's getting right now."

"I know you're right, Elinor — but you've got to realize that he's never bumped up against anyone like you before. For my sake, if not for your own, please cool it."

After one particularly heated altercation over who should pay for my gas and oil while I raced the KR for him, I forgot my promise to Robbie and pointed out that the only reason I hadn't gotten the fuel outright from one of the big companies was that I was too busy keeping flying dates to stop and negotiate for it. Furthermore — and now I was really on a roll — if he didn't consider piloting skills a salable commodity in the job market, how about coming out right now and flying with me? Maybe I could show him more graphically what I was talking about.

By now Robbie was staring despairingly out the window. But Fairchild addressed him sharply.

"Robbie, you know I haven't got time to go roaming about the skies. Pay for whatever she wants." That was Sherman Fairchild.

I studied Butterworth's profile on the drive from the airport to the Edgewater Beach. I could only hope that his offer was a good-will gesture on his part. Up to now the matter of plain unabashed sex had not entered my romantic dreams – and I wasn't about to let it. But I had to work with this man, and I was 800 miles away from the New York NBC headquarters. I'd had little experience in evading this issue since almost all the men surrounding me had been very protective, while simultaneously warning me about the predators among them!

By the time I was installed in a flower-filled suite facing Lake Michigan I'd had ample time to be ashamed of these misgivings, for I had done Butterworth a grave injustice, if only in my thoughts. He told me later it was the only way he could think of at the time to make up for his rudeness on our first meeting. After a day spent behind a roaring engine and signing autographs on a dusty airport, this zephyr-cooled vista was heavenly. In fact, if this was how the other half lived, I seriously considered changing sides.

Luxuriating in a scented bath, I was pulled up sharply by a ringing telephone. (Lady Mary would have loved the Edgewater Beach. All the bathrooms had telephones.) It was Wally telling me to hurry downstairs to meet the NBC staff assembled there. Quite frankly I'd looked forward to dinner on a tray and early bedtime, but by now Wallace Butterworth was batting a thousand in my estimation. If he thought it was important for me to meet the local staff — right on!

Paul Whiteman was the starring attraction at the Edgewater Beach, and Wally was the announcer for his radio broadcast from the ballroom. I had met Whiteman in California the year before, and it was pleasant to renew the acquaintanceship. Also, I loved to dance, and all of the WJZ crew turned out to be just as anxious as I was to get out on the floor. Following the dining and dancing, Wally told me that NBC had exclusive use of one of the chartered speedboats docked in front of the hotel. Heigh-ho and lackaday! While the rest of Chicago sweltered in a midwestern heat wave, we were tracing figure eights on the lake's rippling surface.

On the way to the airport the next day Wally confessed that he was ashamed of the preconceived image of me he'd built up in his mind before my arrival. He also wanted to assure me that while he was responsible to the network for the success of our broadcasts, he wanted to verbally apologize for the strong way he'd come on the day before. From now on it was going to be fifty-fifty all the way.

That certainly made me breathe easier, for I

had a secret plan that would take me away from the microphones, and I might need his cooperation in covering up my absence. I wanted very much to enter the Women's Free-for-All Race to take place on the last day of the air show. Gladys O'Donnell, Mae Haizlip, and Phoebe Omlie all were rumored to be entrants in this race, and I wanted very much to compete against them. I had never flown a race against any females, and now that I had an airplane that was just perfect for this challenge, I felt it high time that I entered the national racing scene.

One look at the opening day's program, and I knew that Wally and I were in luck. I could enter the race, and he'd get to interview some of his heroes and heroines. None other than Lieutenant James Doolittle was scheduled to officially start the air show with an aerobatic demonstration in a Travel Air Mystery ship. Wally gave me the microphone, and I was on the air as Doolittle dove down on the stands until seemingly inches from the ground. He pulled up in a steep 2,000-foot climbing turn, which he spun out of in a series of graceful slow rolls. His aerobatic precision was always a joy to watch, for he made it look so effortless that you could hardly wait to get into the air, hoping that some of his rhythm would rub off on you. It never did, of course, but being able to tell people exactly what Doolittle was doing and why was one of the joys of broadcasting.

He was followed by Marcel Doret of France and Fritz Loese of Germany. I had seen Doret before and more or less knew what to expect, but he surprised us all with a new bag of tricks, including a spectacular loop almost 2,000 feet in depth the nadir of which brought him below the grandstand level. He topped that off with a new version of the three-point landing, by doing them one at a time — first the right wheel, then the left, and finally the tail. The crowds went wild. I was anxious to get Doolittle to say a few words for the WJZ audience and had motioned Butterworth to take over for me when I was handed a note saying that Casey Jones wanted to see me.

After all these months I couldn't imagine why, but — ever the optimist — I hoped that our situation might have improved. The boy who had given me the note motioned for me to follow him, and I found myself trotting back toward Cliff Henderson's office. As I got farther and farther away from the flight line and Doolittle, my good spirits steadily evaporated. This was yet another of fame's unhappy consequences. To refuse the summons invited having it noised around that I'd gone "high-hat."

Yet by accepting it, I laid myself open to exploitation by Casey. What if he wanted to impress someone by telling him how easily he could call me off my job? As I trotted through the darkened hangars, I became more and more angry with myself for having permitted this to happen. He *had* called me off my job — and not only had I let him get away with it, but I'd probably lost for all time the chance to interview Doolittle, who meant more to me than fifty Casey Joneses. ...

I pulled up short as I spotted Casey standing alone in the gloom of an almost empty hangar. I was glad to see he was alone — at least *that* worry was out of the way.

Without so much as saying hello, he demanded, "You still stand in good with Bellanca, don't you?"

"Far as I know, why?"

"I understand George Haldeman's flying the Bellanca Airbus in tomorrow."

"Yes, that's what I heard, too."

"There's going to be a race that she could be entered in, and I'd like to fly her."

"Casey, you wouldn't have a prayer. G.M. wouldn't let anyone fly that creation of his without weeks of careful instruction from George — and as for a race, do you realize how big that bus is? It can seat twelve to fourteen.

He's designed it to corner the transport market. It's the biggest single-engine airplane in existence."

"That's just what I mean," he said eagerly. "I know Bellancas aren't built for speed, but I understand this one steps along pretty good."

"I'm sure she does, but I wouldn't build up any hopes about flying her in the transport race. She'll be George's baby all the way."

"Not if you go to bat for me, she won't."

"Casey, I'm telling you. There is *no* way you are going to get your hands on that cream puff. And as for racing her – forget it." Glancing at my watch, I turned to leave.

"Then you won't intercede for me?"

"Casey, on this one it wouldn't do any good if I tried to intercede for *me*. Anyway, why don't you ask George yourself? You certainly know him well enough."

He looked down shamefacedly – and then I remembered. Back in the days before I made the endurance flight, Casey and his followers were the most vociferous critics of the Bellanca CH monoplane, spreading damaging stories about her so-called instability near the ground.

All the way back to the booth I tried to figure out this strangely insensitive man. Did he think this request was supposed to offer me a way to get back into his good graces? Couldn't be.

Besides, why should I *want* to get into his good graces? Yet by his attitude I could tell that the truth hadn't made the slightest difference. If there was any way to place the blame on me for his failure to race the Airbus, that's exactly where it would go. I couldn't help wondering if George and G.M. planned to race the Airbus. Wouldn't it be great if they did? Just to see that big baby swinging around the pylons would be something at that!

I ducked out to the flight line and, by sheer coincidence, collided with Lieutenant Doolittle. To my joy, he agreed to a radio interview, provided we could do it right away. He was — as always — surrounded by a crowd, so it wasn't easy to open a path to the broadcasting booth, but with the help of some sympathetic onlookers, we managed.

Wally Butterworth almost fell out of the booth in astonishment when we put in an appearance, but he regained his aplomb enough to hand me the microphone and grab a pencil and paper, which I knew would be presented to our guest for an autograph. As for the interview itself, it was a heady few minutes for me and probably a total bore for Doolittle, even though he was much too courteous to say so.

However, after the interview was over, I wasn't too shy to ask his opinion about a problem I was having with the press. As a result of my racing and stunting the Kreider-Reisner during demonstration flights, I was being referred to in the newspapers as a "lady daredevil of the skies." In my view, this was damaging to my image of dependability with the aircraft manufacturers, but I seemed helpless to refute it. Doolittle smiled broadly, saying, "Why, that's a cinch. Just glance at your watch, and shoot your cuffs with confident authority, and when the reporter asks what that was for, just say you're practicing to become a master of the calculated risk! See if that doesn't stop them cold."

After he left, Wally mopped his brow. "Whew! If you know any more like him, don't spring them on me without notice. I've been wanting to shake that guy's hand for the last ten years."

I pretended a yawn. "Stick with me, Wally. You may not wear diamonds, but you'll meet a lot of interesting people!"

Now that I had discovered that Wally was a true flight buff, it was a pleasure to see to it that he eventually got to meet Art Goebel, Al Williams, Roscoe Turner, Gladys and Lloyd O'Donnell, Phoebe Omlie, Wiley Post, Lee Schoenhair, Speed Holman, and a host of others. To return the favor, Wally brought

Ruth Chatterton, Hoot Gibson and his new bride, Sally Eilers, up to the booth. Gibson was a pilot, so we could talk easily, but I left the ladies to Wally.

Gladys O'Donnell won the Pacific Women's Derby, flying the same clipped-wing Waco she'd used the previous year. Well out in front of her competition, she shot across the finish line clocking in at fifteen hours and thirteen minutes for an approximate distance of 2,000 miles. Predictably Phoebe Omlie also won in her Monocoupe for the women's division in the small plane race through the southern states.

The crowds at Chicago were just as big and just as enthusiastic as they had been in Cleveland, but something was different — something I couldn't put my finger on. Even though many familiar faces were absent (Thaden, Earhart, and Lindbergh hadn't come, and Lady Mary still wasn't cleared to fly), I knew that wasn't it. It had to do with a sense of foreboding that I found myself unable to dispel, even though I could find no tangible reason for it. I was shortly to find out the reason for that foreboding.

The army, navy, and marines had turned up with separate stunting teams whose performances were breathtaking, if a bit noisy. I interviewed Lieutenant John De Shazo, one of the navy pilots I'd met at Anacostia during my altitude preparations. De Shazo was a carrier-based pilot from the USS *Lexington* and an entrant in the race for military airplanes that was about to start. We were just going off the air when Al Williams poked his head into the booth, saying to De Shazo, "Come on, boy, you're holding up the show!" and the two of them left arm in arm.

The military race was a grinding fifty-miler. De Shazo fought valiantly on the pylons but couldn't stay in second place. He slipped back gradually but was right on the tail of number two when they flashed over the finish line. The winner, a Lieutenant Cornwall, pulled up in a right roll after crossing the line, and the number two man, Lieutenant Commander Campman, pulled up clear of the field in a straight climb to about 1,500 feet. De Shazo was right on his tail and opted for a left roll to get out of the way of the man behind him. His plane rolled over on its back in a slow and graceful motion – and then dropped like a stone directly in front of the south tier of the bleachers. There was a blinding flash as the gas tank exploded, and a column of black smoke marked the flaming wreckage. De Shazo undoubtedly died instantly, and a spectator drenched with flaming gasoline died hours later in the hospital.

Numbly I handed the microphone to Wally. I could not pretend to be unemotional. I had looked forward to describing the next race because C. B. Allen had finally succumbed to air race fever and was competing against Hoot Gibson and Stud Quimby. But I knew I couldn't handle it after this, so I left the booth. I'd had my eyes glued on De Shazo, so I should have known instantly what would happen. That I didn't indicated that whatever it was had taken only a split second to occur. There were several possibilities. He could have experienced a momentary stall and become disoriented, in which case it would have taken very little pressure on the stick in an inverted position (if he'd pulled it toward him) to drive him into the ground. But this was unlikely for a pilot with his background in spectacular aerobatics. It is more probable that he was buffeted by the prop wash of the plane ahead of him.

By the time I got myself pulled together and back in the broadcasting booth, C. B. Allen had won the race over Gibson, and the crowds were keyed up by the announcement that Art Goebel should be landing any minute. He was competing for first place in the nonstop race from Los Angeles to Chicago. So far Wiley Post was the winner, with Lee Schoenhair in second place. Bill Brock was third with Roscoe Turner – who'd been plagued with engine trouble – limping in to fourth place.

At 5:31 P.M. Art roared across the finish line, nosing out Schoenhair for second place by racking up an elapsed time of nine hours and twenty-one minutes, just twelve minutes behind Wiley, who was now the official winner. Interestingly, every ship in this race was a Lockheed, so the individual scores (with the exception of Roscoe's) were directly attributable to piloting skill.

That night at the hotel I received a phone call from Tim Sullivan telling me I should be prepared to leave Chicago at any time. The takeoff bulletins from Coste and Bellonte were heating up, and this was a landing that NBC planned to pull out all the broadcasting stops for. My heart sank at this news. The Women's Free-for-All race that I wanted to enter was scheduled for September 2, still three days away, and I would have to allow time for flying around thunderstorms that dotted the whole route between Chicago and New York. The only sensible thing to do was to take off the following morning. Once the fliers left France, they'd be bucking the same headwinds that had plagued Kingsford-Smith. However, their biplane, the *Question Mark,* was reputed to be faster than the *Southern Cross,* they just might make it to New York in thirty-six hours or less.

The weather report the next day couldn't have been worse. Thunderheads and even some small tornadoes dotted the route I would have to take. Actually there was no way to avoid them on *any* route to New York unless I detoured over Labrador. I opted to wait one more day. The Lindberghs finally did put in an appearance after all, but this time they just stayed overnight, and the colonel did not participate, as he had in previous years.

The effect of De Shazo's death was sobering, although no more so than the next tragic event. George Fernic, a young Romanian flier, fell to his death in a plane of his own design that he'd been trying to market as the latest thing in aircraft safety. G. M. Bellanca had arrived in Chicago by now, and he witnessed the crash. G.M. knew young Fernic quite well. When Fernic first came to this country he worked for G.M. as an engineer, later going out on his own. According to G.M., a structural failure was the cause of the crack-up.

Three deaths during a single meet was more than I had ever witnessed, and it gave me pause, although like every other pilot there, I had no thought of staying out of the air because of them. Racing *was* dangerous, and attempting an inverted position on a pull-out was something I would never do myself. But I had no doubt that De Shazo had done it many times before with complete success. Fernic was extremely ambitious. I had seen a previous crack-up of his on Roosevelt Field and knew that he would have haughtily rejected suggestions or criticism of his design. Arrogance was dangerous in the aeronautical mileu. Even mild-mannered G.M. had been offended by him. All this I could rationalize. But what of the innocent bystander dying of burns from De Shazo's gasoline? Fate? Destiny?

Crowds are notorious for their short attention spans, and those at Chicago were no different. As I watched them, it was difficult to believe that these disasters had ever occurred. They thrilled to the exploits of Doolittle and Williams and turned away from the parachute jumpers in disinterest once the chutes blossomed open. Only one jumper held their attention over the rest — the delayed drop artist. There was a hush as his plummeting body came into view from his 6,000-foot leap. When his chute opened at less than 1,000 feet with a powerful snap that we later learned had broken his shoulder, they turned away with a regretful sigh.

I thought back over the year before. Despite my two personal tragedies at that time, I had performed daily with no thought of failure. The crowds' cheers rang in my ears then as I basked in their approval. I rarely left the booth now because the autograph seekers hounded me as soon as I set foot outside. I had no desire to perform for them. I could easily have done it. I'd been stunting and racing the KR all summer for Fairchild's Aviation Corporation and NBC. But suddenly the crowds' morbid fascinations became abhorrent. I, who had sought their uproarious acclaim all these years, found myself avoiding it like a discarded lover.

I talked it over with C.B. that night. He regarded me evenly. "You know, there's hope for you yet. I do believe you're growing up. For a while there Deke and I thought it would never happen."

"You're a big help! Here I am trying to figure out what went wrong with me or my chosen field —"

"Hold it right there. Just because you picked this business a couple of years ago doesn't mean it's the Holy Grail, you know."

"Maybe not to you. But I still can't picture myself doing anything else, so why this sudden about-face regarding crowds? I make a large part of my income performing in front of them all the time. It isn't as though I haven't been aware of their attraction to blood and gore all along. I saw enough of that on Curtiss Field as a child. But now I seem to be seeing it with different eyes. I know it's something I've simply got to get over ... but somehow I can't."

"Why not concentrate on broadcasting? You seem to be doing pretty well there."

"True, but at eighteen I was a novelty. I just turned nineteen the other day. That may change the network's attitude."

"Washed up at nineteen, eh?" he teased.

"Oh, no. My contract runs for another year with renewal options after that."

"Well then, I don't see what you're concerned about."

"It's that my future in aviation isn't nearly as clear-cut as I used to think."

"Oh? Whatever happened to becoming the best woman pilot in the country?"

I laughed ruefully. "To tell you the truth, there's much more competition out there than I thought. After watching Thaden and O'Donnell, I realized I'm not nearly so far out in front as I figured."

"But you *are* out in front ..." he said teasingly.

"No, honestly, I'm not all that sure anymore. Even Earhart is catching up, though she's still got a long way to go."

"You're the only one to set four major world's records in the course of a year. Incidentally, are you still keeping track of the number of types and models of aircraft you've soloed?"

"Good Lord, no ... but Mother is, come to think of it. Why?"

"Because if you're really serious about that best pilot business, the American Society for the Promotion of Aviation is making that selection an annual event. They'll be voting on best male and female pilots in the U.S. in about two weeks, and versatility will be important. I don't think any of the others have flown flying boats or trimotors, so you're out in front there, too. Interested?"

"Of course, although common sense tells me not to lose any sleep over it."

"How come?"

"An outfit with a name like that doesn't strike me as being too swift. The whole thing sounds like a PR balloon, and if that's what it is, it will announce the winners as Lindbergh and Earhart, and the rest of us can go back to our aviating chores."

"That's where you're completely off course, my stubborn Irish friend," he said with a boyish grin. "That group happens to be a very prestigious one made up of the top names in the industry. Even G. M. Bellanca belongs to it, you'll be happy to learn, and the voting is done by licensed pilots."

"Hmm, sounds like an impressive vote of confidence in whoever wins."

"You can be sure it will be. Incidentally, have you decided on your route back to New York?"

"Route? You mean skipping between the line squalls, don't you? With all the storms along the way that the weather department knows about, and the ones I'm sure to find that they haven't tracked yet, I'll mow a path over the Alleghenies with my propeller."

"Lots of luck. I take it the next time I hear your dulcet tones, they'll be wafting over the airwaves, telling me about the landing of Coste and Bellonte?"

"Hopefully, yes — provided they make it nonstop. If not, it's back to the old grind, selling White Owl cigars, cold cream, and Coca-Cola."

I left for New York in the face of one of the worst weather reports we'd had to date, but there was no help for it. Coste and Bellonte had already taken off from Paris, so I simply had to get back. Fortunately these cluster storms were of a highly visible nature, and I was able to do exactly what I'd described to C.B. the night before. Instead of one long storm front, they were spaced, presenting the appearance of darkened mushrooms floating over the landscape. I was able to skirt all of them until I approached the lakefront west of Cleveland.

A big black thunderhead menaced the area with lightning bolts flashing from its depths and sheer dark veils of rain reaching to the ground. I tried to keep my distance. I was heading for the Cleveland airport, where I planned to refuel. I'd been battling extreme air turbulence ever since leaving Chicago, and my strength was ebbing, but the knowledge that I'd be on the ground soon was as much of a bracer as I needed.

Suddenly a black mass loomed up in front of me where moments before it had been clear. Without warning the KR and I were sucked up into a nightmare as the rain drenched us and jagged flashes of light illuminated the ground. I fought to stay level, but the altimeter needle was racing around the dial. At one point we were at 9,000 feet, and then, as though an evil giant had tired of toying with us, we were abruptly hurled out of it over the town of Elyria, Ohio. I know it was Elyria because an enormous water tank with the town's name painted on it was directly under me. I have yet to meet anyone from Elyria, but I will never forget that community as long as I live!

The KR and I limped into Cleveland, got gas and the latest weather report, and took off once more. The weather, seemingly tired of bouncing us around, cleared off for the balance of the trip. It seemed no time at all until the welcome sight of Fairchild Field at Farmingdale, Long Island, came into view. I was no sooner on the ground than I was told to take off again. Tim Sullivan had called to say that Coste and Bellonte had been sighted and had radioed ahead that they would land at Valley Stream.

Ah, well, things were picking up all over. . . .

14

Who's the Best Woman Pilot in the United States?

By the time I sideslipped into Curtiss Field at Valley Stream, the weather had changed its mind again. A brand-new storm front moved in from the Atlantic, dumping heavy rain indiscriminately over airborne and land-based spectators alike. Ait traffic was unusually heavy as chartered press planes and Curtiss passenger carriers fought for vantage points from which to view the approach of the French heroes.

To my delight, I'd been assigned Curt Peterson again, along with Kelvin Keech, who would translate our broadcasts for French listeners on a brand-new exclusive NBC international radio hookup. Not being personally acquainted with either Coste or Bellonte, I would have been in a bad way without Keech, for my French was confined to the "please-accept-the-pencil-of-my-aunt" variety taught in Freeport's high school.

Despite the sporadic heavy rains, by six o'clock the field was jammed, and word had been received that the *Question Mark* had just passed over Glen Cove on Long Island's north shore. Curt and I had been on the air for about an hour when I spotted the big red biplane making its approach. The tumultuous crowd held its breath as the *Question Mark* set down in a perfect three-point landing. At that point they went completely mad and surged toward the French airplane, knocking down fences and breaking through police lines in a human tidal wave of hysterical emotion. It had taken three years for Lindbergh's flight to be duplicated in the opposite direction, and the fact that it had taken two men to do it detracted not one whit from their riotous American welcome.

Once again a New York City policeman carried me on his shoulders to the side of the visiting heroes. As the fliers were lifted from their aircraft, exhaustion was clearly etched on their drawn faces. Keech translated into French the questions I fed him, and we learned that it had been anything but an easy flight.

They had followed approximately the same route as Kingsford-Smith and had come up against almost identical weather problems, except for one interesting facet. Owing to a constant battle against ice formations on their wings, at no time had they flown higher than 2,500 feet. When Keech related this, I was awed. It meant that they had not been able to avoid the two most punishing elements on a prolonged over-water flight — turbulence and fog. However, they enjoyed one distinct advantage in the *Question Mark*'s speed. Because it was faster than the *Southern Cross*, they were able to outrun large storm areas of the kind that had threatened to swamp the larger and slower Fokker.

Our interview with them was brief and rewarding, and once again we'd beaten CBS to the punch. As before, CBS announced that Harry Von Zell would handle its microphones and then at the last minute tossed Ted Husing into the fray — but it didn't help. Husing couldn't get to within a city block of the French pilots. Anyway, CBS's equipment broke down again, and it was forced to carry a relayed summation of the fliers' ovation from its New York studio.

The arrival of the Lindberghs at the air field did much to take the crowd's attention away from Coste and Bellonte, giving them a much-needed respite from the adulation of the teeming throng. Both men said over and over to Keech how much they longed for uninterrupted sleep. After thirty-seven action-packed hours at the controls, they were surely entitled to it, but now the problem was to get them off the field at all.

I remembered a back entrance to one of the hangars and told the police, who, by that time, had recovered their equilibrium and taken charge of the two airmen. The cops finally hustled them off to the darkened passageway that led to a ladder taking them up over the hangar tops to the administration building, where they were warmly received by the Lindberghs and the press. We later learned that it had been a difficult day for the French press, who were firmly excluded from this scene by an eager but misinformed Curtiss representative.

NBC's international hookup was operating perfectly, so it was decided to use only two bilingual announcers for the balance of the coverage. I heartily agreed. France's triumph deserved better treatment than those of us limited in linguistic ability were able to give.

I flew the KR back to Farmingdale and listened to the rest of the uproarious welcome at the Ritz Towers Hotel in New York on the radio at home. NBC called the next morning to tell me that I'd been rewarded with a handsome bonus for my work in Chicago and at Valley Stream. Flushed with this victory, I went out to the Aviation Corporation (Fairchild's corporate name) to decide on which model Kreider-Reisner I would use in an aerobatic display Tim Sullivan had booked me into at the county fairgrounds in Danbury, Connecticut.

Robbie came out to the hangar with his familiar worried expression and a large sheaf of papers in his hand.

"Elinor, Mr. Fairchild would like to see you in his office," he said earnestly. Then, dropping his voice to almost a whisper, he asked, "Did you *really* stay at the Edgewater Beach in Chicago and charge it to the company?"

"Uh-huh, why?"

"Mr. Fairchild is very upset. Says you had no right to book yourself into a luxury hotel without prior permission."

"Really? What happened to that part of our agreement that says I am to make appropriate arrangements if my radio work causes any change in schedule?"

"I already pointed that out to him, but he doesn't think this is a reasonable alternative."

By now we were at the great man's door. He didn't look up when we entered — a bit of psychological conditioning not lost on me — but this day I wasn't buying it.

After waiting a respectable few minutes, I took the initiative.

"I understand you are unhappy about my staying at the Edgewater Beach in Chicago."

"I would like to know why it wasn't checked out with this office before you changed your reservations from the Blackstone Hotel," he said stiffly.

"I guess because I didn't understand that I was supposed to. I thought 'making alternative and appropriate arrangements' left any changes in plan up to me. The Edgewater Beach is at least ten miles closer to Curtiss-Reynolds than any of the downtown hotels. Also, I was given a suite for the price of a room. The rest of the NBC crew were quartered there, so the network picked up the tab for my meals and daily transportation to and from the field. I felt that these savings alone would more than compensate Aviation Corporation for the change."

"Hmm, I see — but there is still the matter of the room being almost a hundred dollars more than the Blackstone would have cost."

"Mr. Fairchild, the Blackstone was jammed with air show participants. When I called there to cancel my reservation, the room clerk told me he was glad for my sake because the only room left was an inside single the size of a broom closet. You must have read about the heat wave during the show. It made conditions at the airport comparable to operating in a blast furnace. After a day there, coming back to the noise of roistering pilots and no air conditioning, I would have gotten no sleep. Aside from that, I repeatedly worked the Kreider-Reisner name into both the Chicago broadcasts and yesterday's international radio hookup to Paris for Coste and Bellonte. Frankly there is nowhere else you can get that kind of advertising exposure for a mere one hundred dollars."

He threw up his hands and addressed Robbie. "Are you in cahoots with her?" he demanded. "If I ask any more questions, she'll come up with a whole new flock of answers!"

We had a good laugh, and to all outward appearances, that was the end of it. But I knew Fairchild well enough by now to sense his unease. Ladies in his tightly controlled world didn't make independent decisions or turn up as airplane pilots and announcers in the world of international radio. He was a male chauvinist to his heels, albeit a genteel one. I had every reason to feel that when our agreement ran its course, it would be the last of its kind he would enter into. If he hired a woman at all, she would be docile, undemanding, and cheap. In his book, independence came high, and he would never again contribute to it by paying a single penny more than he had to. I didn't hold that last part against him for a minute. It was only good business. What bothered me was the implication that I was driving a sharp bargain when I made sure he lived up to his end of our agreement.

I left his office and went back to supervise the rigging of the KR. The lakefront storms had given her far more stress and strain than my aerobatics ever would. But there were going to be some racing prizes at Danbury, and I was never one to look the other way when there were a few easy dollars like that lying around. Besides, it would be another way to show off the airplane to advantage. Robbie followed me to tell me that Mr. Fairchild wished me well at the air meet.

"That's nice," I said shortly.

"Aw, come on," said Robbie. "He's not a bad guy, just careful with a buck, is all."

"Robbie," I said wearily, "you're the salt of the earth, and because you are, you've missed the boat entirely. Do you think for one minute that my decision in Chicago would have been questioned had I been a man?"

He shrugged. "I've no idea. But after all, you *are* a nineteen-year-old girl and he's never dealt with anyone of your age-group — or sex — before."

"But that's just the point. He's hired me to do a professional job, which is exactly what I'm doing, plus giving him all this radio advertising for free. He'd be justified in firing me if I didn't deliver. But I *have* delivered, and I resent his reluctance to trust my judgment in expense account areas. I'm no Gene Fowler. I don't believe money should be thrown from the rear of moving trains. On the other hand, I can't do my job if I'm confined to a closet of a room with a naked light bulb hanging from a wire!"

Robbie laughed ruefully. "You sure have a way with words. . . . Anyway, after today's talk I'm sure he sees your point, and I doubt it will happen again. He has friends up in Danbury who've told him they plan to attend, so he'll be anxious to learn how the KR makes out."

At Danbury the fairgrounds were crowded, and the aerial program was long and varied. My aerobatic exhibition was well received, and that was gratifying since I muffed the barrel roll out of the top of the loop (the same maneuver I'd seen Doolittle do so many times) when the motor gave a protesting sputter. I shouldn't have attempted it in front of a crowd because I'd only done it a few times privately and it needed much more working on. But the audience was forgiving or didn't notice (I never found out which). Diving down to pick up flying speed and get the gas flowing smoothly in the engine again, I pulled up into the Smith version of Doret's giant loop — a more modest version, to be sure — but Danbury seemed pleased with it. In the racing program I took two firsts and a second, and then the PA system announced that all transport pilots could carry passengers for hire. Oh, happy day! The last time this had happened, I was winging it on a private license. Now I was *legal,* and it was a thrill to pull up to the flight line and let the potential customers know that I was available.

The little Kreider-Reisner got a workout that day. With riders lined up and waiting, I was kept busy circling the field and stashing money in pockets and paper bags. I was back to my cash-and-carry days of three years before and loving every minute of it. At home in Freeport that night I dumped it all out on the dining-room table and with my race winnings found that I had cleared well over $600 for the day's work!

After I had been back in the New York studio for a week or so, I put in another appearance at Farmingdale. This time I was booked for an appearance at an air show in Red Bank, New Jersey. I was busy with the fueling when Robbie came after me.

"Elinor, have you got time to talk with Mr. Fairchild before you leave?"

"Frankly, no, but if it's important ..."

"Evidently it is to him."

This time my reception was more cordial, but it quickly developed that the boss wanted a run-down on my financial returns from Danbury. I gathered that his friends had reported on the show, and he was most pleased with my aerial performance. But it was clear that he thought Aviation Corporation should share in my profits.

I drew a deep breath, straddled a chair, and sat down.

"Back in the spring, when I gassed up the KR for that appearance down in Charleston, South Carolina, you said that since cross-country trips weren't stipulated in our agreement, the cost of gas and oil was an expense I should personally handle. I was upset then, but on thinking it over, I had to agree that you were right. I had assumed something that I shouldn't have. I took time out to negotiate with Tydol, and they've picked up that tab ever since.

"I don't recall anything in our contract stipulating a split in my exhibition winnings or earnings. So if we are truly operating on a two-way street, I should not have to share either my prize money or passenger-carrying earnings. Am I correct? Or would you prefer to cancel our contract by mutual consent?"

One of America's wealthiest men stared at me wordlessly while Robbie assumed his familiar stance at the window. It was an awkward moment for me and the last straw for Fairchild.

I left his office and told the mechanics to roll the KR back in the hangar, but Robbie came running out to tell me to take off — that everything was straightened out. I knew it wasn't, but it didn't seem fair to disappoint the people in Red Bank.

When I arrived back at dusk, the hangar looked deserted, until I spotted a lone figure lounging in its darkened depths. I taxied inside and shut off the motor. Sherman Fairchild helped me out of the cockpit.

"I waited because I wanted to talk to you alone," he said. "I've thought over what you said today, and I must admit that that part of the contract was never spelled out. I really don't want to cancel our contract unless you do."

Glory be and hallelujah! We shook hands on it, and he walked me to my car. I was five miles down the road before the humor of the situation struck me. One of the most important moral victories of my career had just taken place in an empty hangar with no witnesses, and last week one of my most embarrassing failures had taken place in full view of a packed grandstand. Oh, well, it was a consolation to reflect that it's not every day the boss apologizes to the help.

The publicity and suspense of the national contest for best pilots were building daily in the newspapers, as the society withheld the names of the number one contenders to keep the pot boiling. The day that Bert Acosta and I were locked into second place should have been a red-letter one, but instead, I was encased in a cloud of gloom as I told myself that that was undoubtedly where I was fated to remain. C.B. did his best to shake my conviction that the winners had already been selected and that I would simply have to go back to trying harder. But that was the rub. I'd already tried as hard as I could. There were no more tricks in the bag.

Childhood dreams die hard, and I couldn't bear to see this one bite the dust, especially if the contest wasn't fair and square. To take my mind off it, I called G.M. and started negotiating for a Pacemaker or Skyrocket to boost my own altitude record before anyone else took a crack at it. Also, I just might be able to assuage my bruised ego, which had been buffeted when George Haldeman, years older and many pounds heavier (to say nothing of his superior piloting skill), had passed my 27,418-foot mark by a good half mile, flying the same model airplane. The Wright company had since assured me that it had licked the ice-in-the-gas-lines problem, and I was raring to go.

G.M. was favorable to the idea, but there was a snag. The sale of Pacemakers was moving along so briskly that there would be some delay in setting one aside for this purpose. The Skyrocket was still on the designing boards, getting the wrinkles ironed out of it, so it looked as if next spring were the earliest reasonable target date. From where I sat, that was a long way off, so I suggested having a go at the Airbus. After trimotored Fords and twin-engined Sikorsky amphibians, the single-engined transport should be a snap, and I knew no other female would have the opportunity to fly it. G.M. had invested a great deal in the Airbus, and so far had sold none. We were in the midst of discussing this knotty problem when C.B. called to say that the best pilot contest was all over but the shouting.

"The ballots have been counted, and we're about to run the winners," he said teasingly. I was silent, waiting for the other shoe to drop. "You don't sound interested in who won," he said innocently.

"I just don't want to give you the satisfaction," I said hotly. "Lindbergh and Earhart, right?"

"Wrong, my bullheaded Irish friend. The winners are ... are you ready ... James Doolittle and Elinor Smith!"

"You're kidding ... I don't believe it. Don't tease me about a thing like this, C.B."

"Who's kidding? I just called you for a statement, that's all." I stared at the phone, unable to talk.

"Come on, did you faint or what? I've got to print something. ... You want to give it back?"

"No, no, of course not. But I don't know what to say. You make up something for me. ..."

I walked three feet off the ground for the rest of the day. Having a childhood dream realized is heady enough, but having it happen in the glare of national publicity in tandem with one's personal hero is almost too much.

I thought back on those early years of obsessive practicing. They had certainly paid off, although I was sure I couldn't have gone through them had I known what a long road lay ahead. In striving for the versatility that would assure me employment by the manufacturers of all sizes and types of aircraft, I had consistently hoarded my precious free time and devoted it to further aircraft study.

At that stage in aviation's development each airplane had a set of flight characteristics as distinctive as a set of fingerprints, and the key to obtaining the craft's top performance lay in one's knowledge of these individual peculiarities. One would think that such total application to a subject would inevitably lead to a sense of ennui, but in my case it never did. After a long hard week of wrestling a particularly balky aircraft into usable submission, my idea of a relaxing afternoon was to practice slow rolls in a souped-up Stearman.

Elated as I was to be named the winner, it was almost anticlimactic. The goal was achieved, and right then I wasn't sure why. I would have been devastated if I'd found out I'd been voted for because of the publicity regarding my youth or even as the symbol of the progress being made by women. I yearned to know that the voting was fair and concerned only with piloting ability.

I needn't have worried. A scroll, suitably inscribed, was sent to me a week later with a full-page listing of all the types of aircraft I'd flown and a complete run-down on the speed, endurance, and altitude records I'd set.

Congratulations poured in. I must confess to a deep sense of satisfaction from this overwhelming demonstration of approval from my peers, and it took me awhile to float down from that lofty perch.

When I finally got both feet on the ground, it was a bit of a letdown to find that the earth hadn't wandered out of its orbit a single degree and that while everyone was happy for me, my triumph was not about to change anyone's way of life – including mine.

This was brought home to me when Joe slid out from under Father's Hupmobile and asked me to give him a hand in adjusting its ailing carburetor.

"You mean me, America's Queen of the Air?" I protested in mock dismay.

"You got it, Ace," was his terse rejoinder. "The last time I asked Pop, he got confused, took his foot off the clutch, and almost tore off the garage door. Just pump the gas pedal a couple of times and start her up. If that doesn't do it, we'll have to tear the blamed thing down again."

It didn't do it, and an hour later we both were

covered with grease when Mother called out that Lady Heath was on the phone. I raced for the house. Ever since her release from the hospital, Lady Mary had developed a disturbing pattern of disappearing from view for days, sometimes weeks, at a time. My last letter to her New York apartment had been returned marked "Not at This Address."

Her cheery greeting of "Ma – h – velous, m'deah, simply ma – h – velous!" indicated to me that she was about to pick up just where we'd left off weeks ago and had no intention of disclosing where she'd been or what she'd been doing. I tried repeatedly to find out, but her excited conversational stream flowed on, brooking no interruption.

She was not only ecstatic over my contest win but particularly pleased "that you did it without so much as a single millionaire in sight! We simply must get together soon. I'm thinking of getting married again, and I want you to meet him. Don't tell Willie. He doesn't approve – my age and all. Silly boy . . . one is never too old to be happy, is one?"

No, one isn't, I thought to myself, as I tried to decide how best to return the grease-smeared receiver to its cradle. Lady Heath had hung up before I found out where she was or how to arrange this meeting that she obviously wanted

— but at least she was somewhere close by, and the aviation grapevine would undoubtedly provide the answers.

The coup de grâce to the definition of *best* was decisively delivered by Sonny Harris. I wasn't present at the time, but the scene was described to me later by Emil Burgin.

It was at a hangar flying session that the conversation drifted to how I'd been selected as Best Woman Pilot. One pilot objected to my winning it on the basis of age. At nineteen, wasn't I too young for the title? Sonny glared at him for a moment and said, "Y'know, Sid, that girl has worked harder in her nineteen years than you have up to now, and you're pushing forty. She was born knowing something that you'll never find out."

"Oh, yeah — what's that?"

"That in this world, there's no free lunch — anywhere!"

Epilogue

It seems appropriate to end this story at the point where my impossible dream came true. But there had been so many trials and tribulations along the way that when it finally happened, I distinctly recall a feeling of being washed up on a friendly shore after a shipwreck in a stormy sea. Like all survivors, I briefly gloried in my salvation before discovering that life flowed on around me as it always had.

There was still half a decade of active flying years ahead of me — years in which I would set the world's altitude record for women and a hatful of speed records as I fulfilled contracts for various sponsors. There would be meetings with President Hoover and Will Rogers during which we would hammer out legislation helpful to pilots to this day and a planned transatlantic flight to Rome that took up the better part of eighteen months and then was disappointingly canceled when the great financial depression of the thirties threatened to wipe out my sponsor.

My marriage to Patrick Sullivan, a legislator-attorney, interrupted some of this activity, but it was raising a family of four youngsters that finally brought it to a halt.

A few years after my husband's death, in 1956, I was invited to address the U.S. Air Force Association on Mitchel Field. Then I soon found myself working busily with the Air Force. They didn't have to ask me twice if I would like to fly a jet trainer. Discovering the delights and differences between jet and propeller flying opened up a whole new world. It also led to some hilarious incidents aboard the Air Force's C-119s when I found myself back to dumping jumpers all over the landscape – only this time it wasn't for the newsreels. These men were paratroopers and we were participating in war games.

It was a privilege to fly with this group of World War II pilots and to "hangar fly" with them. What I was able to teach them about civilian parachuting was more than slightly offset for me by their instruction in combat landings. Almost twenty-five years had passed, but they made me feel as if I'd never been away!

At just about this same time in the late fifties, a group of us who had flown on Roosevelt Field thirty years before were brought together by Slim Hennicke and Sonny Harris to form an organization we called the Long Island Early Fliers. Twenty years and hundreds of meetings later my notebooks were filled to bursting as I recorded the tales of old friends and contemporaries. The realization that the development of aviation technology had outstripped all others in the history of man was not lost on these hardy souls; they are all avid fans of today's astronauts.

Much as I agree with them on this score, I still find myself fascinated with the stories and adventures of the gallant men and women I knew so well so long ago. To me it is like looking into an Alice-in-Wonderland mirror whose reflections are standing still in time. Sometimes I have to pinch myself to be sure I am really the girl in those yellowing pictures. At other times I am back in the cockpit of a Waco-10 or the cabin of a Bellanca Skyrocket. This usually happens in a modern airliner as I marvel at its progress over land and sea at 30,000 feet or more.

Surely, there was never a time when I was the only girl in the world to have penetrated these upper regions — or was there?

Glossary

AILERON: A movable section of the wing's trailing edge the function of which is the raising or lowering of the wing during flight.

AIR-MAIL BEACON: A powerful revolving light (much like that used in lighthouses) for guidance of pilots flying the U.S. mail at night.

ALTIMETER: An instrument which indicates the altitude of an airplane over sea level or a set reference level.

BAROGRAPH: A highly sensitive revolving cylinder which records altitude and length of flight. No aeronautical or astronautical records are recognized unless one is installed before takeoff by a Fédération Aéronautique Internationale representative, who also removes it at the termination of the flight. The sealed cylinder is then calibrated at the Bureau of Standards in Washington, and the actual figures are posted. Owing to the barograph's sensitivity, there is usually a higher altimeter reading of at least several hundred feet.

BLIP: To give an engine a sudden short burst of power; usually done to clear out any built-up carbon, but was also sometimes done to clear an obstacle on landing, when prolonged use of power was unnecessary.

CHANDELLE: An abrupt climbing turn preceded by a short dive in order to gain momentum and continue gaining altitude.

DEPARTMENT OF COMMERCE: In the twenties a division of the Department of Agriculture, selected by the federal government as the controlling body for the aviation industry. It was the Federal Aviation Administration of the time.

DRAGGING: Flying over a field at low altitude with full power on in order to make sure the field was suitable for a landing.

EMPENAGE: Airplane's tail assembly.

FAI: Fédération Aéronautique Internationale is the international organization that has the final word on all world's flight records. In the United States it works in conjunction with the National Aeronautic Association.

FIN: Vertical stabilizer positioned forward of the rudder and above the fuselage.

FISHTAILING: Swinging the aircraft violently from side to side prior to making a landing in order to present the broadside of the fuselage to the wind. This has the effect of making the airplane act as its own air brake in cutting the plane's forward speed.

FUSELAGE: The body of the airplane, minus wings and tail section.

HANG ON THE PROP: As close to a vertical climb as the ship will stand and the maintenance of that climb until it starts to stall out.

HANGAR FLYING: Comparable to the golfer's "nineteenth" hole without the liquid refreshments; a time to swap stories and experiences.

JENNY: Also known as a Curtiss JN4-D. A two-place, open-cockpit biplane powered by a 90 h.p. OX-5 engine. Originally designed for World War I and used as a training plane for American and Canadian troops. The U.S. government sold out warehouses full of dismantled Jennies as "war surplus" material after the war at very low prices. Most of the barnstorming Jennies were acquired in this fashion.

LONGERON: One of the main longitudinal supports in the framework of the airplane.

MONOCOQUE: A unique fuselage construction used largely by the Lockheed Corporation. Instead of longitudinal supports covered with linen, the body of the aircraft consisted of a series of "barrel-stave" wooden hoops covered with a thin shell of pressed wood.

OLEO STRUT LANDING GEAR: The early planes all had landing gears with axles between the wheels, which led to many a crack-up. The axle kept the wheels too close to the ground so that even a large clump of grass could cause the pilot to nose over. But the oleo strut gear was almost half again as wide as the axle type, with no obstruction between the wheels. Struts attached to the fuselage were carefully greased and "sleeved," one inside the other, so when the plane landed, the gear spread out and absorbed the shock.

OUTSIDE LOOP: An inverted maneuver during which the wheels of the plane remain in the *inside* of the loop and the pilot's cockpit is constantly on the outside rim. Also, instead of diving down to gain speed and then pulling up to get over on one's back, as in the conventional loop, this one starts with a steep dive beginning at the top with the pilot guiding it around the bottom arc of the loop and climbing out and up on the other side to complete the circle.

PRESSURED EXHALING: In the early oxygen masks it was necessary to pant vigorously and constantly in order to develop one's lung power of suction to draw the oxygen through the feeder tube.

SIDESLIP: To slide an airplane sideways in a downward direction for the purpose of losing altitude without gaining forward speed.

SLIPSTREAM: Turbulence generated by a propeller as the airplane moves through the air.

SQUIRREL CAGE LOOP: An aerobatic maneuver wherein three or more airplanes chase one another around in an enormous loop, several thousand feet deep, like squirrels in a cage. It is extremely difficult to perform since each aircraft is flying directly into the slipstream of the airplane it is following.

STABILIZER: Horizontal airfoil located on the tail. Prime function is to stabilize plane in flight.

TAIL SKID: Flat metal shoe or "runner" to support tail and act as a brake.

VERY PISTOL: A rocket pistol used by ships at sea and early airplanes to communicate distress signals.